MAKER
CITY

Dale Dougherty

Dec. 2016

A PRACTICAL GUIDE TO REINVENTING OUR CITIES

PETER HIRSHBERG

DALE DOUGHERTY

MARCIA KADANOFF

Printed in Canada.

Published by Maker Media, Inc., 1160 Battery Street East, Suite 125, San Francisco, California 94111.

Maker Media books may be purchased for educational, business, or sales promotional use. Online editions are also available for most titles (safaribooksonline.com). For more information, contact our corporate/institutional sales department: 800-998-9938 or corporate@oreilly.com.

Publisher: Roger Stewart
Editor: Marcia Kadanoff
Copy Editor: Mary Doan
Proofreader: Mary Doan
Cover Design: Erik van der Molen
Creative & Book Design: Erik van der Molen

October 2016: First Edition
Revision History for the First Edition
2016-09-15 First Release See oreilly.com/catalog/errata.csp?isbn= 97816804526313 for release details.

978-1-680-45263-1

Maker Media unites, inspires, informs, and entertains a growing community of resourceful people who undertake amazing projects in their backyards, basements, and garages. Maker Media celebrates your right to tweak, hack, and bend any Technology to your will. The Maker Media audience continues to be a growing culture and community that believes in bettering ourselves, our environment, our educational system—our entire world. This is much more than an audience, it's a worldwide movement that Maker Media is leading. We call it the Maker Movement.

How to Contact Us

Please address comments and questions concerning this book to the publisher:
Maker Media, Inc.
1160 Battery Street East, Suite 125
San Francisco, CA 94111
877-306-6253 (in the United States or Canada)
707-639-1355 (international or local)

For more information about Maker Media, visit us online:

→   Make: and Makezine.com: makezine.com

→   Maker Faire: makerfaire.com

→   Maker Shed: makershed.com

To comment or ask technical questions about this book, send email to bookquestions@oreilly.com.

Maker City is a manifesto for productive change in America's cities. With sharp vignettes and instances from around the country, the book depicts a new groundswell of entrepreneurial and civic-minded energy that is bringing tech people and crafts people together to work—city by city—to create new partnerships to reinvigorate city-based industries.

Amy Liu

Vice President and Director, Metropolitan Policy Program, The Brookings Institution

WASHINGTON, D.C.

This is a wonderfully specific and useful book about one of the most promising economic and social trends in America. The positive prospects that Peter Hirshberg, Dale Dougherty, and Marcia Kadanoff lay out closely match what I have seen in cities across the country. And the spirit of resilient adaptability that they portray, including practical steps on fostering it, will be an important part in the next stage of U.S. economic growth.

James Fallows

The Atlantic

WASHINGTON, D.C.

Maker City shows with clarity and rich examples ways in which cities can harness the tremendous potential of the Maker movement and urban manufacturing to create a more resilient local economy and more vibrant city.

Kate Sofis

Co-Founder of the Urban Manufacturing Alliance

SAN FRANCISCO, CALIFORNIA

Maker City is a must read for anyone concerned with the economic future of their community, city, or nation. I believe that this is one of the most important urban planning books since Jane Jacobs' The Death and Life of American Cities. It compels, no ... demands that we all go to work on the future of our cities. The Maker City is an achievable vision. If you care about the future or want to help create it ... read this book.

Mark Hatch

Co-Founder, Former CEO, TechShop

SAN FRANCISCO, CALIFORNIA

Begins the conversation about how to remake our cities for the better in a way that sidesteps partisan politics. Reading this I'm optimistic for what our cities could be, thanks to the Maker movement.

John Clippinger

Author of From Bitcoin to Burning Man and Beyond: The Quest for Identity and Autonomy in a Digital Society, MIT Media Lab

CAMBRIDGE, MASSACHUSETTS

Chock full of examples of how Maker Cities are getting cities right in ways that are clever, inspiring, highly replicable, and sustainable, not faddish.

Hugh Mackworth

former venture capitalist, tech executive, and father of a 16-year old Maker

PORTLAND, OREGON

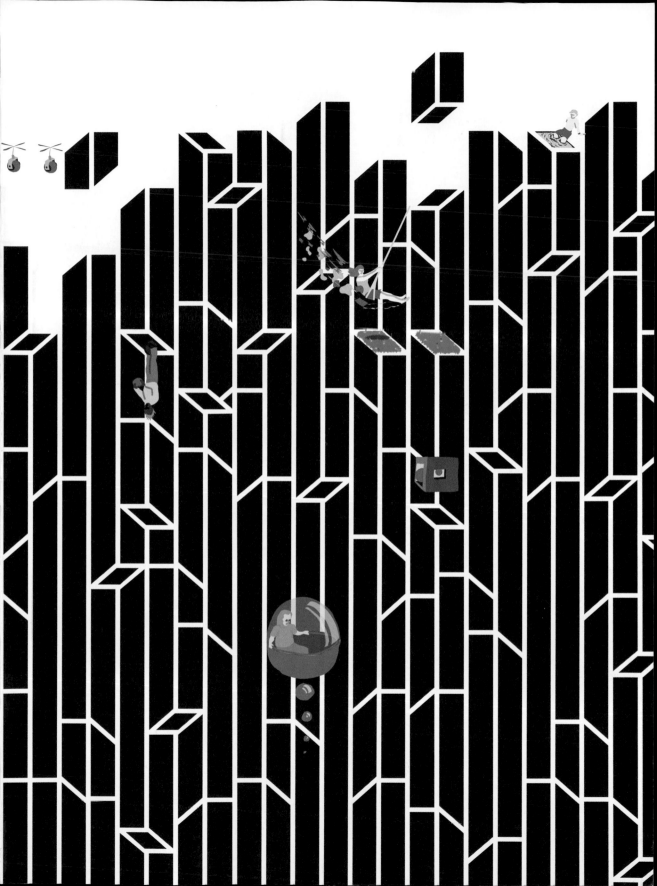

Table of Contents

BUILDING A NATION OF MAKERS

BY: THOMAS KALIL

Thomas Khalil is Deputy
Director for Policy for the
White House, Office of
Science and Technology
Policy, Executive Office of
the President; Senior Advisor
for Science, Technology and
Innovation for the National
Economic Council

Recently, the Office of Science and Technology Policy hosted a White House meeting of over 175 organizers of Makerspaces. Passionate and creative organizers shared their ideas for building vibrant and inclusive Maker communities in cities and towns across the country.

The meeting supported President Obama's **Nation of Makers** initiative, which he launched at the first ever White House Maker Faire in 2014. This effort is designed to ensure that more students, entrepreneurs, and Americans of all backgrounds have access to a new class of technologies—such as 3D printers, laser cutters, and desktop machine tools—that are enabling more Americans to design, build, and manufacture just about anything. This initiative was featured in a report of **100 examples** of President Obama's leadership in science, technology, and innovation.

WHITE HOUSE | NATION OF MAKERS
mcbook.me/2cqDjV7

WHITE HOUSE | 100 EXAMPLES
mcbook.me/2cXT9Gi

In recent years, participation in the Maker Movement by individuals and organizations across the United States and around the world has grown dramatically. More individuals are interested in being producers, not just consumers, and 2.3 million people have attended Maker events around the world. The hardware and software tools needed to design and make just about anything are becoming more powerful, less expensive, and easier to use, and individuals can now get access to these tools at both commercial and nonprofit Makerspaces, TechShops, and Fab Labs. More and more organizations are recognizing the value of Making for education, workforce development, innovation and entrepreneurship, advanced manufacturing, and economic development, including leading companies such as Intel, Autodesk, Chevron, Ford, Google, and GE.

President Obama has recognized the Maker Movement as a national priority for a variety of reasons.

First, Making advances values and disposition that are ends in themselves, such as curiosity, creative confidence, self-expression, invention, and collaboration.

Second, Making has a role to play in education and lifelong learning. It can inspire more young people to excel in STEM (science, technology, engineering and math) subjects, connect their learning to real-world, personally meaningful problems, and reverse the decline in student engagement. Survey data reveals that two-thirds of high school students report being bored every day.

In 2010, a team of 15 teens from a low-income school in West Philly showed us what's possible when our young people are challenged to solve real-world problems. When competing for the $10 million Automotive X Prize, they built a fuel-efficient hybrid car that outperformed other fuel-efficient cars built by professional engineers and graduate students from top universities. In a region with a dropout rate of over 50 percent, every single member of the team graduated. Inspired in part by that experience, the teacher who led the team has now launched an entire public school focused on student learning through solving hands-on, real-world problems.

Third, Making could help promote innovation in hardware and manufactured products. In the same way that cloud computing and open source software has lowered the costs and the barriers to entry for digital innovation, Makerspaces could make it easier for an entrepreneur with a great idea for a manufactured product. Entrepreneurs can also take advantage of crowdfunding, open source hardware, accelerators and incubators focused on hardware startups, and low-cost components (e.g. sensors, cameras, semiconductors) created for smartphones. In the same way that companies such as Apple emerged from the Homebrew Computer Club, the 1970s version of the Maker Movement, some Makers are already turning "pro" and launching startups.

Finally, Makers can become an enormous asset for solving societal challenges. One nonprofit (Tikkun Olama Maker) brings together Makers and people with disabilities to co-design assistive technologies. These "Makeathons" have created prototypes of solutions that allow people with disabilities to get in and out of a wheelchair, open doors, grab objects, and go kayaking.

If we want to ensure that more Americans have access to these opportunities, we need to take the phrase "Maker Movement" literally. Movements have goals, and the organizational capacity to mobilize people and resources to meet those goals.

One concrete next step would be increase the number of regions that have mapped their assets and current initiatives, identified an ambitious but achievable goal, and are bringing together the people, organizations, and resources needed to achieve that goal.

Fortunately, a growing number of organizations are recognizing that they can both benefit from and contribute to the Maker Movement.

→ Companies such as Ford Motor are serving as "anchor tenants" for TechShops. In addition to helping Ford increase the number of employee inventions, this TechShop also benefits the broader Detroit community.

→ Schools such as Elizabeth Forward in western Pennsylvania have built a "Dream Factory" – a Makerspace that allows students to work on real-world projects that integrate Career and Technical Education, computer science, and the arts. This has increased student test scores and graduation rates. School leaders in 1,400 schools representing 1 million students have signed the "Maker Promise" to unleash students' passion and capacity to make.

→ Case Western Reserve University is investing in a 50,000 square foot facility called the Think[box] with space for prototyping, fabrication and business incubation. MIT is allowing students applying to MIT to submit their Maker portfolio in addition to their SAT and GPA scores.

→ Over 100 Mayors have taken the Mayors' Maker Challenge because of its potential to foster economic development, job creation, and entrepreneurship in advanced manufacturing and hardware. Many cities are participating in the Urban Manufacturing Alliance, which is pulling together experts in workforce development, real estate development and local branding. These efforts are helping entrepreneurs make the transition from Maker to manufacturer.

→ Libraries, science museums, after-school programs, and other nonprofits are increasing access to Makerspaces and mentors. In Pittsburgh, foundations such as the Grable Foundation have helped create a network of 250 organizations (the Remake Learning Network) committed to reimagining learning for the 21st century. This network recently announced commitments of more than $25 million to support hands-on, personalized learning, including Making.

The world needs more Makers, inventors, problem-solvers, and entrepreneurs. In the years ahead, I hope that more individuals and organizations work together in communities across the country to increase the number of people – young and old – that have the chance to get involved. For, as President Obama noted, we must "recommit to sparking the creative confidence of all Americans and to giving them the skills, mentors, and resources they need to harness their passion and tackle some of our planet's greatest challenges."

This preface was originally published by the Huffington Post under the title "Building a Nation of Makers".

AN INTRODUCTION & CALL TO ACTION

OUR GOAL FOR THIS BOOK IS TO SHARE THE INSIGHTS, KNOWLEDGE, AND PRACTICES OF HOW PEOPLE AND ORGANIZATIONS INSIDE AMERICAN CITIES ARE LEVERAGING THE MAKER MOVEMENT TO:

→ **build community**

→ **create economic opportunity**

→ **revitalize manufacturing and supply chains**

→ **reshape education and workforce development**

→ **redefine civic engagement**

This book is for Makers, community organizers, business leaders, policy makers, and other city leaders in local government.

The ideas presented in this book can apply to any city or town. What matters is not the size of your city or the availability of specific resources but the level of engagement of your residents and citizens.

As you read this book, we hope you will:

1. Reconsider what "progress" looks like.

Progress used to be synonymous with what was big or built for a mass market. This put pressure on our cities to get rid of factories, because they were dirty, dangerous, and too big and sprawling to make for good neighbors. As a result of this pressure, we ended up outsourcing much of our manufacturing capacity and prowess to other countries, particularly those with inexpensive labor and lax regulations around the environment and workforce safety.

Today, we see that consumer taste has changed, moving away from products produced for a mass market to those that can be:

→   produced cost effectively in smaller lots

→   designed as platforms to be hacked by the consumer

→   changed frequently in response to consumer feedback and improvements in technology inputs

Together, these trends have redefined what progress means. Instead of thinking big and building for a mass market, think small and building to iterate for agility.

All of this favors tighter supply chains and placing manufacturing in close proximity to product development. Suddenly the future looks like many smaller Makers and manufacturers joining together into a network, like the internet itself.

2. **Think of a Maker City as a way to be resilient in the face of technological change.** At a time of great change that raises questions about the very nature of work, a Maker City is how a community prepares for the future. A Maker City provides access to tools in a much more democratic manner, so that all of its residents are able to learn new skills, express themselves, and become more entrepreneurial and future oriented.

The Maker movement is comfortable reconciling two great forces in our society: it both embraces the latest in technology and simultaneously asks what this means for people and the definition of meaningful work. Maker Cities are living experiments. Their learnings and successes suggest the contours of a hopeful economic future.

There is great concern in our nation that perhaps many jobs will be replaced by robots. The Maker movement provides another lens into this future, asking the question: "How do we enable people to reach their full potential and do work that is additive to our society, leaving the routine and mind-numbing work to robots?"

How a city frames questions about and organizes its responses to technology change has everything to do with its future and its resilience.

3. **Understand that the Maker City has many experiments and many voices.** This book is about the many experiments and voices in the U.S. at work on reinventing our cities. And our future. We encourage you to add your voice by posting Maker City stories on Twitter and Medium using the hashtag #makercity.

Cities are finding that the rise of the Maker movement is a transformative moment, one they can seize upon to create jobs and economic opportunities, bring manufacturing back into urban settings, and reshape education and workforce development. Ecosystems and real estate are being developed inside cities to support Making.

The Maker City isn't just a place for new forms of economic production, it's about the Maker ethos, which combines a can-do spirit with a sense of agency.

Ultimately, the Maker City is a federated story and one that hews close to deeply held American values of invention and collaboration; you won't find a Makerspace where people aren't collaborating or teaching one another.

4. **Recognize this is a time for optimism and opportunity for U.S. cities.** The Maker movement is reshaping U.S. cities, creating an entirely new wave of economic opportunity, one that has the potential to involve people from all walks of life and life stages and do it in a way that is uniquely American. To paraphrase President Barack Obama, the Maker movement is inventing the jobs and industries that will define our future:

President Barack Obama at the first-ever White House Maker Faire, June 18, 2014

"Our parents and our grandparents created the world's largest economy and strongest middle class not by buying stuff but by building stuff — by making stuff, by tinkering and inventing and building; by making and selling things first in a growing national market and then in an international market — stuff 'Made in America.' ... Your projects are examples of a revolution that's taking place in American manufacturing — a revolution that can help us create new jobs and industries for decades to come."

HOW TO TELL IF YOUR CITY IS A MAKER CITY?

One question we get asked a lot is how to tell if a city is a Maker City. It is, if you want it to be. In other words, being a Maker City is an aspirational goal.

There is an undeniable level of optimism, energy, and economic opportunity that is evident every time we speak with someone who lives, works, or makes within a Maker City. Over 100 cities in the U.S. identified themselves as Maker Cities, as part of a White House initiative to build a **nation of makers**.

Maker Cities embrace not just the Maker movement but also the tenets of **Open Innovation**, the idea that the best solutions do not lie with any one individual or institution inside city government but must be created through collaboration and engagement that looks outside for answers and examples of what to do to affect change.

Today, cities in the U.S. have a unique opportunity to more broadly engage with their citizens and residents in developing creative solutions for some of our most pressing urban problems Problems such as building affordable housing, eliminating homelessness, reducing congestion on our streets, eliminating gun violence, keeping older workers engaged and productive across a longer life expectancy, and ensuring that every child has the opportunity to succeed first in school and then in the world of work.

The Maker movement encourages people from all sectors (public, private, academia, non profit) to come together and try to solve the problems that a city faces in a collaborative and open fashion. In a Maker City all of these problems and more are being addressed by Makers, entrepreneurs, and open innovation.

Jane Jacobs, the noted urbanist, writing in her landmark book The Death and Life of Great American Cities, had this to say:

Cities have the capability of providing something for everybody, only because, and only when, they are created by everybody.

WHITE HOUSE | NATION OF MAKERS
mcbook.me/2cqDjV7

OPEN INNOVATION
mcbook.me/2cquBD8

HOW TO BUILD
A MAKER CITY

Ten first steps any city can undertake:

1. **Maker Best Practices.** Lead or participate in local efforts to identify, document, and share "promising practices" in manufacturing and technological innovation so that others in your community and beyond can learn from local experimentation.

2. **Maker Liaison.** Designate a Maker liaison in the mayor's office or economic development department.

3. **Maker Roundtable.** Host a roundtable in your community that convenes partners and helps catalyze public and private commitments that will strengthen the local Maker movement.

4. **Maker Faire.** Help celebrate the ingenuity and creativity of local Makers by holding or participating in a Maker Faire event which convenes stakeholders to promote innovative technology ideas. See MakerFaire.com.

5. **Makerspaces.** Host or help create or grow Makerspaces in local incubators, accelerators, educational institutions, under-utilized buildings, and/or design-production districts which can broaden access to tools needed for design, prototyping, manufacturing, and the growth of small business enterprises that are building new manufacturing and innovation technologies.

6. **Maker Manual.** Issue a "Maker Manual" to explain the importance of the Maker movement in your community and to identify resources and incentives at local,

regional, state, and national levels that can support
Makers and small businesses seeking to grow their
technology and manufacturing innovations.

7. **Make a Strategy for Education, Training, and Workforce
Development.** Commit to working with your school
district, libraries, museums, after-school providers,
community colleges, universities, workforce investment
boards, and job training organizations to give more
students access to age-appropriate Makerspaces
and mentorships, and focus more education and
training programs on the emerging fields of advanced
manufacturing and technology innovation.

8. **Maker Business Development.** Upgrade your economic
and business development programs, incentives,
and services to provide support to manufacturing
entrepreneurs and small businesses.

9. **Maker Support in Struggling Neighborhoods.** Support
initiatives to engage and support students, entrepreneurs,
and small businesses in underserved neighborhoods.

10. **Make It Even Better.** Your community may have even
more innovative strategies for promoting the Maker
movement. Make that part of your Challenge pledge
and share the strategy with others! A history of
technology innovation is not required to be a Maker
City. Maker Cities are rewriting the narrative around
technology, recognizing that not every city can or should
remake itself in the image of San Francisco, where it
seems there is a technology startup on every corner.
In the Maker City, technology is an enabler, one that
gives Makers new tools they can work with to build
advanced materials into their products, to support rapid
prototyping, and to resurrect urban manufacturing.

KAUFFMAN FOUNDATION
| MAYORS ACTION
REPORT

mcbook.me/2clS8UJ

In the 1960s this begged the question: how do you get everybody involved in something as complex as a city? The Maker movement has opened up the exciting possibility within our cities to engage more people in the process of co-creation, both in solving the problems that living in a city presents and also in co-creating the institutions that define a city.

INNOVATION: ONE OF MANY DRIVERS OF RE-URBANIZATION

The U.S., like the rest of the world, is re-urbanizing at a rapid rate. **By the year 2050, 70 percent of the planet's population is expected to live in cities.**[3]

The landscape of business invention or more broadly, innovation, has changed over time. From the '50s through the '80s, many major businesses that relied on knowledge and tech workers moved to the suburbs to create corporate campuses. The idea was that companies could assemble all the resources they needed "to be innovative" in one place, and put them on a closed campus so that employees could work together. The assumption here was that the best ideas were the ones that didn't leak out. Of course, innovation wasn't likely to seep in either.

The trend has shifted in recent years, exemplified by the mid-Market Street area in San Francisco, where within a five-block area tech startups and the infrastructure they depend upon for growth are located in close proximity to each other. This encourages cross pollination across the startup ecosystem that exists in San Francisco, all the better to encourage open innovation. Density is the new incubator of innovation.

GE is a company that is ranked #8 on the Fortune 500 list and is known for its innovative nature. Recently, the company decided to move its corporate headquarters from suburban Connecticut to the Boston waterfront, to better expose GE's executives to a "sea of ideas" around the Internet of Things.

According to Jeff Immelt, CEO of GE:

"I want people who are down in the Seaport, I want them to walk out of our office every day and be terrified. I want to be in the sea of ideas so paranoia reigns supreme. To look out the window and see deer running ... I don't care about [that]." (Source: **Boston Globe** [4], 2016)

Together, demographic changes and changing preferences make for ballooning populations inside our cities that test the limits of infrastructure, particularly around transportation and housing.

The Economic Impact of the Maker City

Given these dynamics, it is apparent that we must get cities right or we will fail as a species.

Maker Cities are getting cities right in ways that are clever, inspiring, highly replicable, and sustainable, not faddish.

The economic value of one Maker City hotspot, the Brooklyn Navy Yard, has been estimated at $392M in direct earnings from the 10,350 jobs created and $1.93B in economic value. In Portland, the local Maker economy has had significant impact: 120 manufacturers created 1,000 jobs, resulting in $270M in local manufacturing revenue.

Additional data on the economic impact of the Maker City can be found in the Appendix to this book.

OUR VISIT TO PROVIDENCE, RHODE ISLAND

We visited AS220, a Makerspace started by Bert Crenca in Providence, Rhode Island. Providence is like a lot of college towns on the East Coast. Every year, its population of 177,000 swells as students flow into Brown, Johnson and Wales, Providence College, Rhode Island College, and Rhode Island School of Design (RISD).

Historically, Providence was a mill town that produced textiles and jewelry. Today the mills are silent and these industries are all but gone, lost to China and other countries where the labor is cheap and manufacturing is set up to mass produce goods. Providence was hit hard by the recession of 2008–2009 and never really recovered. Poverty is high, particularly among **communities of color**[5]: 20.6 percent of Blacks and 31.0 percent of Latinos in Rhode Island live in poverty, according to the 2014 census.

When we arrived at AS220, a 16-year old young man took the initiative to show us around and make introductions. We asked him, "How did you get involved in all this?" He told us, "I used to get in trouble. A lot of trouble. I wasn't doing too well. Then a friend brought me over, and now I'm here all the time. I have more self esteem than I ever imagined." He could have been right out of central casting for the screenplay with the story arc: "troubled youth transformed."

The young people who the AS220 Youth Studio targets have rarely had an easy time in life. Many come from impoverished neighbor-hoods, some have failed in school while others are in the care or custody of the state. About a third of the 450 young people whom AS220 Youth serves each year were previously incarcerated at the Rhode Island Training School (RITS). Makers offer workshops on such topics as creative writing, photography, hip-hop, and design to help them discover new talents, while forming relationships with adult role models. Making is "a way of connecting with very, very disconnected, hard-to-engage young people in a way that doesn't feel like a program with adults telling them what to do," Program Director Anne Kugler states.

AS220 encourages young people to come to its downtown Providence Makerspace to learn new skills, master the tools of Making, and create the kind of portfolio that leads to jobs.

Electronics and the basics of programming are taught at AS220 Industries through a unique partnership with a company called Modern Device. Generally, this type of activity starts with a champion in the community; here the champion was Paul Badger, who teaches a class at RISD the college calls "Physical Computing."

We might call this Maker 101: how to use a low-cost microcontroller and the Arduino development environment to control something you've made. Arduino boards are able to read inputs—lights on a sensor, a finger on a button, or a Twitter message—and turn them into an output: activating a motor, turning on an LED, or publishing something online. You can tell your board what to do by sending a set of instructions to the microcontroller on the board.

Today, every dollar that Paul earns selling electronic kits through his website Modern Device goes to AS220 Industries to enable underprivileged kids to learn the basic skills of "physical computing," skills that the staff at AS220 hopes will lead to jobs in urban manufacturing, robotics, and the like.

This is an example of a Maker (Paul) taking something he's made and leveraging it to change his city for the better.

AS220's task is not an easy one. "It takes a lot of heart to do the work we do," Kugler admits, noting that despite its best efforts, AS220 can't make a difference for everyone who comes through its doors. Yet, for many young people, AS220 is, as one young man put it, "a place like no other." Another participant said of AS220's supportive, inspiring environment: "I think it's a family. I think it's a movement. I think it's a remedy." (Source: **National Arts and Humanities Youth Program Awards 2012**[6])

FIGURING OUT WHAT YOUR CITY IS GOOD AT

Each city has its own unique characteristics and strengths. Instead of trying to recreate places like San Francisco or Brooklyn, the Maker Cities we met as part of this book are building out strong Maker ecosystems by identifying their competitive advantages and leveraging them.

The typical way to evaluate your strengths as a city is by doing a SWOT analysis, but there are other **frameworks**[7] you may find valuable, as described by the Cities Alliance, a nonprofit group which focuses on cities and the eradication of poverty.

What's critically important here is to understand your competitive strengths as a city. What is your city's founding story? What assets are unique to your city that you can leverage? Does your city have a history of urban manufacturing that you can work to rekindle? Do you have colleges and universities that are creating a talent pool unique to your city? If so, how will you get them to stay after graduation?

For example, because Providence is home to both RISD (strength: design) and Brown (strength: engineering) and has inexpensive space (former mills), it is in an ideal position to create hardware startups that want or need to manufacture their product locally. We see that Pawtucket—a former mill city that sits just adjacent to Providence—is becoming a hotbed for Makers who came to Providence for college but remained after they graduated, going on to establish workshops, ateliers, and manufacturing concerns at places like the Ocean State Maker Mill.

Focus on how to build a unique value proposition for your city based on its strengths, starting with the Makers who, in all likelihood, already exist in your city. Examples of unique value propositions for cities are available from the **Kauffman Foundation**[8] which also publishes data on startups and entrepreneurship that can help you understand where your city sits relative to other **cities**[9].

ABOUT THIS BOOK

This book came together through the dedication and efforts of Peter Hirshberg (civic innovator and entrepreneur) and Dale Dougherty (godfather of all things Making).

After attending the 2015 White House Maker Faire, Peter and Dale decided to create this book to answer the questions they were getting from city officials about how to harness the Maker movement and turn it into a groundswell of economic opportunity inside their cities.

This question was one of particular interest to the Ewing Marion Kauffman Foundation which provided funding for this book. The

White House Office of Science, Technology, Policy (OSTP) was instrumental in guiding our thinking around this topic and also in introducing us to many of the Makers we interviewed as part of this book.

Marcia Kadanoff (civic innovator) joined the team as co-author, and a long list of contributors and reviewers got tapped as well.

TO HELP YOU UNDERSTAND AND ORCHESTRATE THE MAKER MOVEMENT IN YOUR CITY, WE'VE ORGANIZED THIS BOOK INTO EIGHT MAIN CHAPTERS:

→ **Chapter 2—The Maker Movement and Cities**
Setting the table with concepts essential to the Maker movement; what is Making all about, the Maker ethos, the Maker mindset.

→ **Chapter 3—Maker Ecosystem**
The ingredients to consider when building and activating an ecosystem around Making in the Maker City.

→ **Chapter 4—Education**
How the Maker movement is remaking the learning community, both informal and formal, from K-12 through adult education.

→ **Chapter 5—Workforce/Economic Development**
How to turn the Maker City into a magnet for talent by turning traditional notions of workforce development upside down.

THE MAKER MOVEMENT AND CITIES

A SOCIAL MOVEMENT THAT IMPACTS BUSINESS, EDUCATION, AND CULTURE

IN THIS CHAPTER, WE DISCUSS WHAT MAKING IS ALL ABOUT, THE TOOLS THAT MAKERS RELY UPON, THE MAKER MINDSET, AND THE MAKER MOVEMENT.

As long as there have been cities, there have been Makers. In cities, people have traditionally made things that other people needed and sold those things in local markets. Yet, in an age of iPhones designed in California and made in China, and when Amazon and Walmart dominate the distribution of physical goods, it's easy to ignore where things are made and where we buy them. Making is returning to cities in new and surprising ways as a growing number of people identify as Makers. Making has become a social movement that impacts business, education, and culture.

Today's Makers are crafters, artists and artisans, technologists, hobbyists, amateur scientists, entrepreneurs, engineers, woodworkers, roboticists, and many others. They are young people engaging in hands-on projects that introduce them to science and technology in creative ways. Makers are also adults who see themselves as inventors and experimentalists. Some have PhDs and others are self-educated. Makers are practicing a craft or challenging themselves to learn a new hobby. They are creative problem-solvers who gain the confidence that they can tackle ever-larger problems.

Increasingly, one of the problems Makers are tackling is how to remake the city they live in for the better. In this chapter, we set the stage for the chapters that follow, by focusing on how Makers create value through collaborative work, leverage open innovation to create business value, work to create meaningful products through informal collaboration, and zone in on "small" as the unit of change.

MAKING CREATES VALUE

Today, there are new ways to make things which means there are new ways to create value. Making as value creation can take place almost anywhere but increasingly, it is happening as a productive, collaborative activity in cities. This value might be personal, as a form of self-expression. The value can be social, in how it connects us to our family and community. Making also has educational value, offering a context for experiential learning and developing self-directed learners.

Increasingly, Making creates business value, as Makers take their products to market and set up manufacturing operations inside U.S. cities.

Today digital technologies, new tools, and new institutions and business models are greatly accelerating the "democratization" of manufacturing. The Maker movement is moving beyond its artisanal and hobbyist roots to embrace the creation of business concerns. Small and lean, but serious, startups are evermore possible given cloud-based tools, new kinds of contractors, specialized studios, incubators, and accelerators.

New technologies (virtual-to-real design tech, cloud computing, cheap sensors, iot) and new resources (manufacturing-oriented incubators, hackerspaces, and boutique contract manufacturers) are making it possible for a manufacturing startup to occur and grow. All of this is quite consequential to cities.

Making has Business Value

The global Maker community can be seen as an open innovation ecosystem based on open source principles. It creates new ways for people to work together beyond geographic or organizational boundaries. Hal Varian, Chief Economist of Google, calls it "combinatorial innovation." People build on the work of each other and more easily discover and utilize resources that are needed to develop new products and businesses. Crowdfunding is one example of community-supported innovation. It creates an accessible path to the formation of small businesses. Increasingly, larger companies are understanding how to collaborate and co-create with the Maker community to extend their own R&D efforts.

Venkat Venkatakrishnan, Director of Research & Development for GE Appliances in Louisville, Kentucky, is behind the development of FirstBuild as an online platform and an open physical space for development of new product ideas for household appliances.

"At FirstBuild, we not only design products but also build products and sell products. We do all three" said Venkat. "We are now convinced that the future of product development will be similar to the way we are doing things."

Venkat Venkatakrishnan
in front of FirstBuild

We'll return to Venkat and FirstBuild when we discuss ecosystems (Chapter 3) and also urban manufacturing (Chapter 6).

MAKING IS A MINDSET AS MUCH AS A TOOLSET

While this mindset develops from the creative and collaborative practice of Making, it fosters a sense of agency, resourcefulness, and resilience. Makers reflect what is traditionally called "a can-do spirit." Throughout our work on this book, we saw Makers organize around a set of common values that lead to community building and civic engagement. The Maker City is a mechanism to spread a participatory, problem-solving culture at a time when it is needed most.

Making requires a place to Make, typically a workshop or studio. More formalized and sometimes larger spaces are called "Makerspaces" and typically include space to work as well as access to equipment such as 3D printers and CNC machine tools. Makerspaces do not just provide space; they can also function the way social clubs did for previous generations, bringing together people with shared purpose and values.

Louis Mumford wrote that "a city is a theater for social interaction." Today we might say that a Maker City is a workshop for social interaction.

In Makerspaces, Makers have access to the means of production at a very low cost.

It's never been Easier to Realize a Crazy Idea and Hold the Result in Your Hand

Makers are generously sharing their skill and expertise with others, in schools and libraries, and in the community at large. Makerspaces offer workshops and courses to help others develop as Makers. Some of the Makerspaces in the community are freestanding; others are part of K-12 schools, community colleges, and universities, as well as museums and libraries. It is important that these places exist, but even more important that they connect to each other, often informally, and foster collaboration and shared resources. Events such as Maker Faires, hackathons, meetups, and markets invite more people to participate and get involved. The Maker movement is about more than just making. It's about the voluntary association, cooperation, and shared purpose that cultivates a local Maker culture. It can come to define a city. A Maker City.

Part of the power of the Maker movement is that it provides a framework for understanding the past, as well as the present and future of a city. Most cities have an "origin story" based on what they made. Trenton, New Jersey's slogan was: "Trenton Makes. The World Takes."

While many Rust Belt cities have suffered from the loss of heavy industry, it remains a source of pride that is still reflected in the character of the city and the people who live there. Making can re-open the discussion about what's made in a city and how that becomes part of its present and future identity. It goes beyond the products and services offered locally and speaks also to the shared values of people in the community.

Macon, Georgia has a population of 150,000. Nadia Osman serves as Director of the makerspace called SparkMacon. She came to this position having previously served as the Director of Revitalization and Business Alliance for the College Hill Alliance, a nonprofit organization tasked with revitalizing a section of Macon that lay between Mercer University and the city's downtown.

"If I really want to get someone more traditional to understand what this Maker movement can mean I say, 'Do you remember when things were made in the U.S., wasn't that cool?' And they say, 'Yes, is that what you're trying to get back to?' And I say, 'Yes.' And that's a whole other spark that we've created."

Nadia also hopes that the Maker movement helps more people in Macon become comfortable doing new things that are non-traditional.

Manufacturing is a traditional business that has declined in the U.S. as more manufacturing has gone offshore to China, Mexico, or Vietnam. The Maker movement has provided a way to re-think how we make things and is reinventing manufacturing in the U.S. by placing it at the crossroads of the creative and tech industries.

THE MAKER MOVEMENT STARTED AS A PROTOTYPING REVOLUTION

Focus on Meaningful Products

New tools, which are now more powerful and more accessible than ever, combine software and hardware and allow more people to use more machines more easily. The internet also connects machines and the things made by machines. There are two resultant impacts. One is that the cycle time between design and prototyping is radically reduced. So called "rapid prototyping" means that product design can happen faster and with greater flexibility, making it more responsive to change. The second is that the ability to design and build prototypes is not limited to industrial designers or engineers. Not only is the threshold lowered for expertise, so too is the amount of money required for parts and/or for facilities.

Nonetheless, there remains a significant gap in the expertise and resources required to move from prototype to production. Often a product needs to be redesigned to take advantage of higher volume production methods. But these higher-volume production methods may not be a good fit for today's demand curves, with a focus on more customized products.

Makers intent on bringing a product to market may not have the capital it takes to set up and pay for high-volume production facilities. Recognizing this, companies as big and global as Flex are stepping in to create new forms of factories designed to solve this problem by supporting short production runs where quantities can be adjusted gradually as more demand for the product is evident. We'll discuss this more in Chapter 6 on Advanced Manufacturing.

Products Need to be Designed to be Manufactured Locally

Kate Sofis is the founder of SFMade, an organization with 600 members that connects Makers to manufacturing resources throughout the U.S. Kate believes that one of the most important trends impacting cities is what she calls "design for local manufacturing."

"If product designers and developers had knowledge of the local manufacturing resources and supply chain, then it might influence how they designed the product. In many cases, thinking about manufacturing happens once a product is already designed. A consumer product that needs to be produced in very large volumes at the lowest possible price to be competitive is, in effect, a product designed for China. However, there is a new design space that optimizes for other factors, such as customization."

Kate explains:
"When the customer has a role in specifying a customizable element of the product, that is probably the number one kind of product that cuts across different types of manufactured products that we're seeing done in the U.S. The classic example in San Francisco is Timbuk2. They are a perfect example because they do both production of bags overseas, and they do production of bags here in the U.S. About 40 percent of their production is out of the factory in San Francisco. And it's a 100 percent customizable product.

Timbuk2 customized bag

"What's interesting to me now is that it used to be just the customizable kind of bag. You'd order it online and it was mostly, well exclusively, for domestic customers. Now they have a very ingenious strategy where they're setting up new retail stores in different locations both in the U.S., and now abroad. They opened their first overseas store in Singapore last year. And the whole gist of those stores are they're basically customization labs with limited inventory."

INFORMAL COLLABORATION

At a foundational level, a Maker City offers many ways for people to get together and generate a broad range of collaborations. Meetups and unconferences are two ways that people meet each other based on their interests.

Nick Pinkston is CEO of Plethora, a company we'll talk about later in this book in Chapter 6 on Advanced Manufacturing. When Nick first moved to San Francisco, he didn't know as many Makers as he did in his hometown of Pittsburgh. He began organizing the Hardware Startup Meetup. Typically, these informal gatherings took place in a warehouse and attracted one to two hundred people monthly. Makers would take turns giving two-minute pitches to talk about a startup or a product that they were working on. Nick said he became known as the hardware guy and people came to him. "That's how I got a really good network," he said. The Hardware Startup Meetups have spread to dozens of cities, creating connections and providing momentum to those who want to turn their ideas into a hardware product or service.

ORD Camp started as a long weekend gathering at the Chicago offices of Google. It was not organized by Google but Google provided space, as one of the organizers worked for Google. It is a type of "unconference," meaning that there is no fixed program prepared in advance. The people who are invited to the event are expected to create the program once they arrive, creating talks and discussions around their interests, however far ranging they might be. While it started with a core tech audience, ORD Camp has expanded to include more and more interesting people from around

Chicago. ORD Camp has created a Chicago-based network of people and resources that is centered on Chicago but not limited to it. Its mailing list of ORD Campers is active, providing connections, advice, and support on everything from local accountants to diversity training. ORD Camp is not a business, and nobody really owns it. It is a loose set of fairly inexpensive resources and events that keep a community together and growing.

Maker Faires

Maker Faires bring together Makers with community stakeholders in business and education in a city. In 2015, there were 90 Maker Faires in the U.S. and 61 outside the U.S. A variety of organizations host Maker Faires, showing that it isn't just one type or organization that plays a leading role.

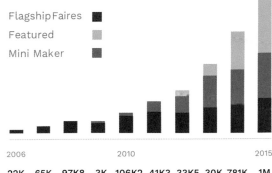

Numbr of maker faire attendees by year

Flagship Faires
Featured
Mini Maker

2006   2010   2015

22K   65K   97K8   3K   106K2   41K3   33K5   30K   781K   1M

MAKER FAIRE [1]
makerfaire.com

Case Study: How Maker Activity Came to Flourish in San Diego

Travis Good is a tech executive who became fascinated by the Maker movement and the spread of Makerspaces in the United States. He was one of the founders of NoVa Labs in Reston, VA. When he moved to San Diego, he began looking for the local Maker community and ending up producing Maker Faires in San Diego. Over several years he worked with a variety of institutions that align around a particular shared vision and a set of principles like collaboration and openness.

"When we did our first Maker Faire, it would never have happened if a company like Qualcomm hadn't signed on to be our premiere sponsor and fund the thing. Their CTO has been a Maker forever and understood its importance. The second Maker Faire was much more broadly supported by the community and other institutions, including the following:

Maker Faire San Diego

→ Office of Mayor Faulconer

→ Balboa Park Museums, ten in all but Fleet Science Museum foremost

→ Makerspaces: FabLab San Diego and Open Source Maker Labs

→ San Diego Central Library and its branches

→ K-12 schools. A network of schools called High Tech High and other project-based learning schools

→ Universities such as University of California, San Diego (UCSD),University of San Diego (USD), and San Diego State University (SDSU)

→ City College, which is reinvigorating itself through their Making programs

"The Dean of the Engineering School at UCSD now has a require-ment where every single year for each of their four years of undergraduate studies, students are required to take at least one hands-on project-based course. It was not at all a requirement previously. And you see the same sort of thing surfacing at USD and SDSU, so the universities here are getting it.

"There are places in San Diego, such as the Oceans Discovery Institute, the new Children's Discovery Museum, and the Elementary Sciences Institute, which are all very much engaged in trying to help the underserved with STEM subjects. They are reaching out for help. Funding is always an issue for this sort of a thing, but it's how, as we become a more tightly knit, better-coordinated community, they can become a part of it, and start benefiting.

"There's a lot of that going on here. It's just that it needs a whole lot more of the rest of us to help it be successful."

EMBRACING CHANGE

The Maker City is a strategy for embracing change not just by adapting to the future, but also by helping to co-create it through collaboration. A Maker City gets started by grassroots-level self-organizing activities that generate a lot of informal collaborations. Many of these collaborations are started informally by individuals, based on their own interests and do not necessarily represent their institutions. It can lead to more structured collaborations such as the Remake Learning Network in Pittsburgh with over 250 organizations and partners working together to create more engaging learning experiences for students.

There is a rather popular view that change happens in big ways, from gaining consensus and moving forward together, often with a central authority as the manager of that change. All resources and efforts are combined to create change. This typically means change is initiated in a top-down manner. However, the Maker movement is characterized by bottom-up, cooperative efforts. Many are succeeding by starting small and allowing small changes to be replicated across a network. People organizing a new Makerspace learn from (and replicate) the work that others have done to start a Makerspace. Makers starting a crowdfunding campaign learn from previous campaigns. The same is true of Makers trying to produce a product or run a business.

Improving the accessibility of tools and resources and the connectedness of members in the community is important for developing a Maker City.

LOCAL AND GLOBAL

A Maker City is connected globally and locally. It's not only one or the other. It's both.

While the Maker movement is global in reach through the internet, it also demonstrates the value of local initiatives, such as community Makerspaces where people meet face to face. They are places to go and meet people, to share ideas and expertise, and to generate new opportunities. While one is no longer limited to local resources, they remain important because that's where we most easily find people.

Making change is more easily done locally. Whether it's starting a business, renovating a building, or organizing a meetup, all depend on organizing local resources. While a business depends on local resources, it also needs access to manufacturing and distribution for global markets.

Case Study: Mokena, Illinois. A Labor Union and Makerspace Work Together

Mokena, Illinois has a population of 15,000 and is located just south of Chicago.

Spacelab, is the Makerspace in Mokena with a lofty mission: We're building the infrastructure for more kids and adults to connect to a future in which they can personally change, modify, or 'hack' the physical world, creating things that were nearly impossible to do on their own just a few years ago.

Jay Margalus is one of the founders of Spacelab and a full-time professor at DePaul University with a research interest in big data and video game design. He says that Spacelab "reflects what our values are and how we best espouse them." He sees Spacelab as a learning community for Mokena, one that hosts workshops but also goes out into schools to run programs.

To start the new space in 2015, Jay ran a Kickstarter campaign, one of the first to raise money on Kickstarter to support the launch of a Makerspace. "It's not really about the money but it was an opportunity to ask the community for support," said Jay. They earned $4K from their campaign. "It's in an old building built for the telephone company," he said. The Makerspace has about 30 members and also has co-working space.

In 2015, Jay and other members decided to organize the Southland Mini Maker Faire approaching the city for help with finding a location. Jim Richmond, an elected trustee of the city, made the introduction to Pipe Fitters Training Center Local 597. Jim, along with Jay, met with Kevin Lakomiak who is charge of the apprentice programs at the facility.

It can be difficult to explain the Maker movement to someone who doesn't know what it is.

"We mentioned the word 'drones' and I saw that we hit a nerve," said Jim. Kevin spoke up and said: "We've been thinking about how to use drones."

That led to an interesting conversation about how drones might be used to inspect the top of a water tower to examine a problem before putting someone at risk climbing up it without knowing what the problem was or the tools needed to fix it. They agreed to explore it further, which led to them connecting Kevin with one of the local Makers who was into drones. "We were offered the union facility for free to put on our Mini Maker Faire," said Jay.

"Pipefitters do a demanding job," said Jay. "They need a large training facility for the different kinds of work they do." The training center has mockups for specific training in a hospital, on an oil rig, or at a waste treatment facility. Jay felt that the union and the Spacelab shared the same values. "My father was a union electrician," he said, adding that his father was the source of his own values. Jim said the people they worked with at the union had the same mindset and he was interested in getting union members to showcase their work. "They do electronics and welding as part of their job, but they also do stuff in their garages or in groups—they are tinkering all the time." He liked seeing them show their work in public and "bringing them into the fold" of the Maker movement.

The collaboration between Spacelab (and three other area Makerspaces) and the Pipe Fitters Local 597 was positive and included other local organizations such as the Boy Scouts. "We had the right mix of people with the right attitudes," said Jim. "Nobody was too big for their britches."

IMPLICATIONS FOR CITIES

Sociologists talk about the strength of ties between people, distinguishing between people you know well ("strong ties") and those you don't ("weak ties"). Experience tells us that many of the best examples of innovation around Making happened when people with "weak ties" get together to share experiences around Making.

To strengthen these weak ties advocates for the Maker movement should work to:

 Establish space and regular venues for like-minded Makers to get together, particularly those who don't know each other well. This can be at a Makerspace (of course) but other venues are also important. Consider hosting your next Maker Meetup at a local community college, the YMCA, JCC, Boys and Girls Club, a union hall, or inside a corporate innovation center. The important thing is to share your passion for Making with others, ideally by meeting them where they already spend time.

Cadence matters. Give people an opportunity to get together one time per month, at a minimum.

Be inclusive on purpose. While young people are an important source of energy and new ideas for the Maker movement, we also value people who have mastered the trades of woodworking, metalworking, or welding over a course of 10, 20, even 30 years. Gene Sherman is CEO of Vocademy, a Makerspace based in Riverside, California focused on practical training; he calls his instructors "masters'" and values them for their gray hair.

When you plan meetups and other ways to bring Makers together, don't focus exclusively on technology or the latest, greatest 3D printer or CNC (computer-numerically controlled) machine tool. Makers need an opportunity to figure out how the tools apply in a real world way. Go back and look at the discussion between Jim and Kevin if you doubt this. Here were two Makers who couldn't figure out how to connect to each other until they discovered a shared passion for drones. Drones, bots and robotics, custom prosthesis for humans and pets, advanced manufacturing ... these are just some of the applications of Making we are seeing inside our cities.

Don't forget city officials. Up until now, Maker Cities have been built through informal and grassroots efforts. Reach out to city officials who have shown an interest in Making and encourage them to be a catalyst for additional Maker activity in your city. Your city could even consider creating a summer internship or other position in city government around Making, so as to encourage entrepreneurial activity, urban manufacturing, and/or workforce development.

There is considerable crossover between the tech community and Makers. If you are having difficulty finding people and support for Making in your city, look no further than the technology company in your backyard. Likewise, look at educators inside your schools, colleges, and universities.

THE MAKER CITY AS OPEN ECOSYSTEM

The Maker City is a dynamic, open ecosystem of resources that spur economic and cultural growth through collaboration and innovation. An ecosystem is no more and no less than a loose coalition of people and organizations that share both an interest in something and a similar set of values.

Almost every city we talked to when developing this book had an ecosystem they could point to although some didn't use that word.

The Maker ecosystems we've seen start out in a self-organized fashion when like-minded individuals with a shared sense of purpose connect people, ideas and projects together in a city. Seldom is any one person or organization "in charge." Leaders emerge who can greatly influence and accelerate the growth of the ecosystem but they might be anyone; a specific job title or function is less important than passion, enthusiasm, and energy.

All cities have the potential to build a rich ecosystem around Making for their economic benefit. Start by understanding the fundamental nature of your city and build from there.

UNDERSTANDING ECOSYSTEMS AS CENTER AND EDGE PHENOMENA

Throughout this project we worked with Deloitte, a consulting firm widely recognized for its ability to recognize the trends that shape our future. Under the leadership of John Hagel, John Seeley Brown, and Duleesha Kulasooriya, Deloitte's Center for the Edge studies innovation and how emergent forms of change diffuse through institutions and society.

In 2013, Deloitte's Center for the Edge worked with Maker Media to publish the **first-ever study**[1] of the impact of the Maker movement on the U.S. economy.

Why is an Ecosystem Important?
A key tenet that runs throughout the work done at Deloitte's Center for the Edge is that activity on the edge of an ecosystem is a precursor of change at its center.

Think of it this way. When activity starts at the edge of an ecosystem, it marks the beginning of change. As the ecosystem grows in the number of connections and the strength of the ties between players, the change starts to make its way into the center. In other words, a strong Maker ecosystem can affect change at a number of levels within cities that, from the outside at least, look impervious to change.

In this chapter, we'll discuss how a Maker ecosystem gets started, how it manifests itself in Makerspaces and Maker Faires, and how these activities at the edge have the kind of ripple effect that over time, changes the fundamental nature of the Maker City at its very core. Most of the cities we spoke to considered the formal learning community to be at the core of the Maker ecosystem. It's not good or bad to be at the core versus the edge; these designations are more about how long the institution has been in place within the city and therefore about the institution's openness to innovation and rapid change.

Often we find that the Maker ecosystem in a city starts at the edge, reaching out and into the learning community, and then growing up and out in an organic way to business and industry, eventually engaging the core where local and regional government agencies sit.

There are two ways to maximize change, according to Deloitte. You can let the ecosystem self organize or you can set up an organization or person to orchestrate interactions. What doesn't seem to work during times of rapid change and innovation are highly centralized structures, what Deloitte calls "hub-and-spoke" structures. These tend to limit interactions between participants in a way that is not beneficial and should be avoided.

Activation is Important

Deloitte and others who study ecosystems understand that it is not enough to declare that you have an ecosystem; you must also encourage people and organizations to exchange and interact. This process is often called "activation."

Makers tend to activate around the resources they have access to, such as a physical space like a Makerspace; parks or public squares where their prototypes get displayed; tools, like 3D printers or sewing machines; events, festivals, and Maker Faires; and funding opportunities.

Activation also gets kindled by local community members, with the availability of space playing an important role, as we'll see in the case studies that follow.

The strongest Maker Cities organize around nodes and networks that allow anyone to become a part of and contribute to the new ecosystem.

UNDERSTANDING THE MAKER ECOSYSTEM

Find out about hardware Meetups at meetup.com.

Maker Cities vary in terms of the strength of their fundamental ecosystems around Making. Those with the strongest ecosystems activate nodes at the edge as well as in the center.

At the edge, the Maker City Ecosystem is made up of:

→ **Makerspaces, fablabs, and fabrication shops.**

→ **Makers as individuals and groups.**

→ **Activist communities.** Community organizers with a specific interest in creating economic opportunity for young people are often a great starting place.

→ **Artist communities and industrial art centers.**

→ **Faith-based communities.** A good starting point is with a Youth Minister, Assistant Rabbi, Arts Coordinator, or Community Outreach Specialist.

→ **Manufacturers and Suppliers.**

→ **Tech communities.** Search "Making," "robotics," and "drones". Seek out startup companies and entrepreneurs.

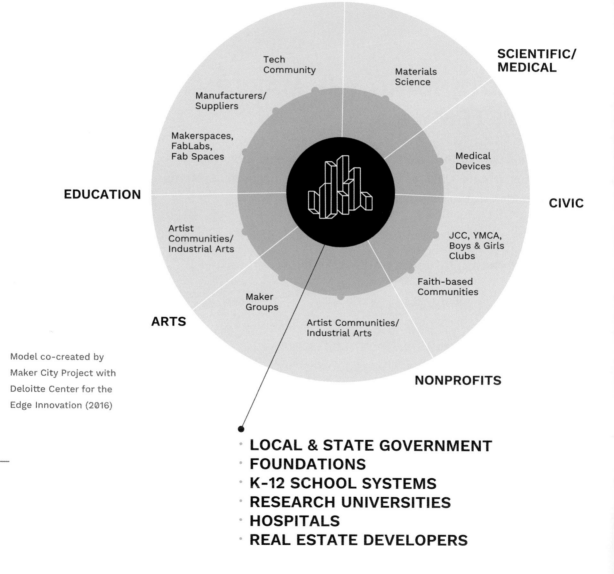

BUSINESS/
MANUFACTURING

SCIENTIFIC/
MEDICAL

CIVIC

NONPROFITS

ARTS

EDUCATION

Tech
Community

Materials
Science

Manufacturers/
Suppliers

Makerspaces,
FabLabs,
Fab Spaces

Medical
Devices

Artist
Communities/
Industrial Arts

JCC, YMCA,
Boys & Girls
Clubs

Faith-based
Communities

Maker
Groups

Artist Communities/
Industrial Arts

Model co-created by
Maker City Project with
Deloitte Center for the
Edge Innovation (2016)

· **LOCAL & STATE GOVERNMENT**
· **FOUNDATIONS**
· **K-12 SCHOOL SYSTEMS**
· **RESEARCH UNIVERSITIES**
· **HOSPITALS**
· **REAL ESTATE DEVELOPERS**

Pittsburgh

The Gable Foundation engaged libraries, museums, nonprofits, and Carnegie Mellon University in re-imagining education, ultimately bringing about change in multiple school districts.

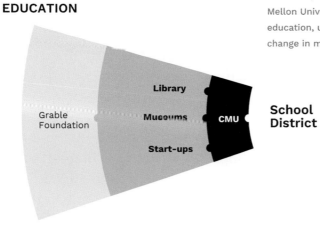

EDUCATION

Grable Foundation

Library

Museums

CMU

Start-ups

School District

ARTS

San Francisco

An experiment which originated in arts and urban design organizations engaged real estate developers and foundations, ultimately bringing about change in the San Francisco Planning Department.

ARTS

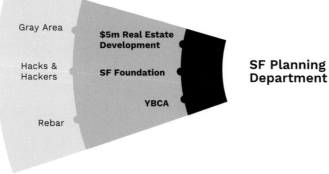

Gray Area

$5m Real Estate Development

Hacks & Hackers

SF Foundation

YBCA

Rebar

SF Planning Department

NONPROFITS

At the center the Maker City Ecosystem is made up of:

→ **Local and regional government.** Engage with your Mayor, the Chief Innovation Officer (if your city has one), and/or the Office of Economic Development.

→ **Corporations.** Work with Community Development officers, the Chief Innovation Officer, the Director of Innovation, the General Counsel's office in charge of IP, or anyone who has spoken on behalf of a company for Open Innovation, as well as the philanthropic arm of larger companies such as Salesforce.org.

→ **Real Estate Developers.** Look for real estate developers in your city who work with mission-oriented nonprofits. Also find developers who have an interest in the adaptive reuse of older buildings and/or who are committed to helping young people advance in your community.

→ **Philanthropic Foundations.**

→ **Research Universities.**

→ **Hospitals.**

→ **Nonprofits.**

In between the center and the edge, the Maker City ecosystem is made up of:

→ **K-12 Schools, Community Colleges, and Universities.** Target science teachers, principals, department heads especially in STEM/STEAM, businesses, and designers who have shown an interest in Making.

→ **Libraries and Museums.** Librarians are often huge champions for the **Maker movement**[2]. Also check out your local science or art museum.

- → **Recreation/Community Centers.** Maker programs are appearing in after-school programs, at rec centers, and inside parks.

- → **Community-based organizations.** YMCAs, JCCs, and Boys & Girls Clubs are obvious places to start.

- → **Design and Engineering Firms.**

It is particularly important to engage people from the tech industry, associated creative industries, and manufacturing industries, as well as selective representatives from the scientific and medical communities.

Artisanal and craft businesses are also important to include. Additionally, cultural organizations in art and music, organized as collectives or shared studios, should be included.

MAKERSPACES PLAY A PIVOTAL ROLE IN FORMING THE ECOSYSTEM AROUND MAKING

Makerspaces can reside in schools, libraries, or museums, or can be larger and more freestanding. What's important is that every city have a dedicated space for Making where the relevant equipment is available, not just for young people but for others in your city interested in using the tools of Making to create businesses and economic opportunities.

Creating a Makerspace need not be expensive. You'll read examples here of Makerspaces that got started on as little as $5K. Today, that is not the norm, largely because Makers have grown in both their levels of sophistication and expectations.

MIT created **FabLab**[3] as a lightweight and inexpensive lab that could be built inside schools and junior colleges. Today, there are

1,000 FabLabs in 78 countries, most costing $25K-$65K in capital equipment to start up and about $15K-40K in consumables per year.

Artisan Asylum (2014)[4] provides a list of starter equipment for a Makerspace and estimates the startup costs at $40K-$70K.

Space is also important for Makers. We see some of the bigger and more elaborate freestanding Makerspaces being created to house not just the equipment of Making but also co-working space, space to host meetups, and other convenings. Operational costs can exceed $1M a year for this kind of Makerspace, many of which double as innovation centers in their communities.

An innovation center is a large building or complex of buildings that can serve as a hub for entrepreneurial activity in your city. Innovation centers can be modeled along the lines of the Cambridge Innovation Center, the largest such space in the country, which provides co-working space for startups. Others are more membership driven, designed to make the tools of Making available for a low monthly fee and offering classes that build Maker skills, much like a gym. Ford and TechShop just collaborated on a Makerspace that is 33,000 feet large, membership driven, and designed to serve as an innovation center for the City of Detroit. We discuss this more in Chapter 5 when we discuss Workforce Development.

PUT YOUR CITY ON THE MAP

The Maker Map is a crowd-sourced directory of the physical locations of Maker resources. You can use it to discover the resources that are already known in your area, such as Makerspaces, manufacturing resources, and retail businesses.

The Maker Map is a great example of the self-organizing principle. It is most comprehensive in cities that have literally "put themselves on the map."

Creating a map of the Maker resources in your community can be a powerful way to connect people together.

If the Maker Map doesn't adequately reflect the resources in your community, then organize a meeting of people to gather that information collectively and put it into the Map. The Maker Map shows how technology can play an important role in making resources more visible and, as a result, more accessible.

3D Hubs is a distributed 3D printing service that allows its customers to use a local resource to get a 3D-printed object. Its map identifies individuals or services that participate in its network and provides a view into the local Maker community. New York leads the nation in the number of 3D hubs with nearly 500 3D hubs in their network; Los Angeles is next with 381 hubs.

Another example of an excellent Maker Map comes from Shenzhen, China. Created by Seeed Studio for Maker tourists, **The Shenzhen Map for Makers**[5] lists factories, Makerspaces, leisure activities, and electronics markets for shopping.

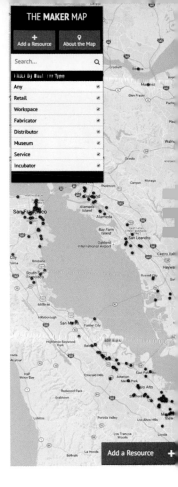

An example of The Maker Map

SURVEY THE MAKER ECOSYSTEM

After a map, the second step is to survey Makers in your city so as to better understand what they are doing and the kind of support Makers need. Portland is a great example.

Portland's Artisanal Economy

Portland, Oregon takes pride in the artisans, crafters, and Makers who work there and who are supported by the many people who choose to buy locally-made products. It's a place that draws talents from other parts of the country for its tech and creative industries, yet it also attracts the independent-minded who want to work for themselves and are looking to create meaning and purpose in what they do.

Portland's ADX, founded in 2011 by Kelley Roy, has become a hub for the local Maker community. "ADX is a collaborative Makerspace where individuals and organizations make and learn. By sharing tools, knowledge, space, and experience, we are doing things better by working together," explains the website. Roy created not only a space but also developed the Portland Made Collective to represent

An example of 3D Hubs map

the people and their products from Portland. Roy, in her book Portland Made, says that she "hoped to create a place for creative people from all industries to work ... and turn those ideas into businesses." There are almost 400 members of the Portland Made Collective, an impressive array of artisanal, small batch producers.

In 2014, Portland State University Professor Charles Heyring surveyed 126 members of the Portland Made Collective (PMC) and **published a report**[6] stating that the members "employ an estimated 1,024 persons and generate revenues of $258M." In this survey, members reported "very positive revenue growth with an average of 61 percent cumulative for the last three years." Members also preferred to identify themselves as Makers or artisans rather than entrepreneurs or business people.

Respondents were asked to identify where their sales came from. The report found a fairly even split between local and non-local sales.

"Given the size of most firms, their products, and their assumed preferences for everything local, we expected a larger percentage of reported sales to be local. While the share of Portland Metro sales was substantial (46%), an even larger share (54%) of sales were outside the Portland Metro area. It is noteworthy that 30 percent of U.S. sales were beyond the Northwest region and that international sales accounted for eight percent of total sales."

A conclusion of the report is that cities should support the growth of smaller enterprises, many of which are still proving themselves:

"As expected, most enterprises are quite young, with 62 percent in operation five years or less and 83 percent ten years or less. It is noteworthy that the three enterprises that have been in operation for thirty years or more produced 90 percent of the revenues and 70 percent of the jobs. The lesson is not to ignore the smaller enterprises but to nourish them. Two of these three large companies were started in small studios by founders with a passion for their work and the ability to turn that passion into something substantial."

In 2014, the Portland Made Collective (PMC) had 126 members. By 2015, the number of members had increased 2.7x to 342 members. **The 2015 survey of the PMC membership**[7] estimates revenues of $316M. Across all categories, PMC found revenue growth to be a very healthy 37 percent from 2014 to 2015

The survey also found that clustering Maker enterprises is a critical for Makers and Maker-enablers as well as those who seek their products. "In interviews, Makers identified proximate access to the Maker ecosystem as important for inspiration, problem solving, resource exchange, marketing opportunities, and a way to benefit from the collectively created Portland Made identity."

Fielding a local survey of Makers is an important step in learning more about who they are, what they do, and how a Maker City can best support them. On the last two pages of the 2014 Portland Made Report you'll find a **sample survey**[8] you can work with and adapt it for your own purposes.

ENGAGE EDUCATORS

Remake Learning Network, Pittsburgh, PA

Pittsburgh, Pennsylvania is a Maker City with a strong Maker ecosystem. This wasn't always so. The ecosystem that developed there came from the realization that the education system was failing its children. Gregg Behr was early in his tenure as Executive Director of the Grable Foundation when he interviewed stakeholders in education:

"The thing that I heard them repeat that dumbstruck me was this, that 'I'm not connecting to the kids the way that I used to.' On one hand, that was totally normal because I'm sure generations of adults have said that about kids today.

"But what was so striking about it was that they literally meant kids from 2005 to 2006, or kids from 2005 to 2007. So for us, that presented the question: is there something seismic that's happening among youth and youth culture, in a way that should prompt us to think differently about how we're supporting schools, museums, libraries, and then the construction, the exhibit and the design of camps, as well as the professional development to support all of these types of institutions that could or should be relevant to kids' lives?

"Simultaneous to that happening, I had occasion to have a simple coffee meeting with a woman who had her own gaming company in town but was also an adjunct professor at Carnegie Mellon Entertainment Technology Center. And when I met with her she turned my eyes open to a whole group, a cadre of roboticists, gamers, and technologists of all sorts who also were thinking differently about their products and services, and how they were supporting kids and education and learning.

"... I thought, if early adopters in the education field, and this whole other crowd of people are thinking along similar lines, could we bring them together? So, I literally organized a pancake breakfast, there were probably about 10 people there, who themselves were from different backgrounds. There was a roboticist there. There was a teacher there.

"You had about a dozen people, and it was just one of those moments. It turned out to be this enriching conversation, and people said this is the new narrative about learning. And I could imagine two or three other people who I heard bring this conversation. So, that's when, at the time it was called, we ultimately called it The Kids+Creativity Network, and over the past few years, it's been re-named to The Remake Learning Network."

Today, the Remake Learning Network has broad community support and is "highly coordinated communities of people and organizations allowing for synergies, swapping ideas, rubbing shoulders, and working together toward a shared vision."

Sunanna Chand, Remake Learning's Impact Strategist explained: "Since its establishment, the network has grown to over 250 organizations, including early learning centers and schools, museums and libraries, after school programs and community nonprofits, colleges and universities, ed-tech startups and major employers, and philanthropies and civic leaders. Together they have created more than 100 Makerspaces throughout the country."

THE ROLE OF CORPORATIONS IN BUILDING OUT THE MAKER ECOSYSTEM

Louisville, Kentucky and GE Firstbuild

Corporations can play a leading role as facilitators of a Maker City ecosystem, connecting not only with Makers but also with local government and universities.

Across the United States, the rate of innovation in product development life cycles has impacted how local economies can adapt to market dynamics and demands. The products we once bought and kept for a lifetime are now being disrupted by information technologies. Just as consumers expect continual upgrades on our mobile phones, the same is true for durable goods like appliances and cars. This is proving to be a significant challenge for cities whose economies and cultures are built on traditional industries. Cities are now tasked with reinventing these industries to stay competitive in both attracting talent and large companies.

Looking to leverage entrepreneurial networks to help them become more innovative and competitive, GE decided to establish FirstBuild, a Makerspace micro factory and cocreation community for household appliances. GE's Appliance Park is located in Louisville, Kentucky but GE wasn't sure there was enough talent in the city to make the project successful.

They were originally looking at cities in the Sun Belt or on the East Coast. City officials wanted FirstBuild to locate in Louisville and organized a meetup at the local Makerspace, LVL1, which had established a thriving community of Makers.

Much to GE's surprise, when they presented their ideas to local Makers, the interest was high and the questions indicated that Makers already had ideas about what they'd like to do to create smart appliances.

When GE opened FirstBuild, they organized community hackathons and noted that the GE employees of Appliance Park were participating on their own time in significant numbers. Louisville's commitment to innovation, its history in manufacturing, and the growing Maker community that was already enmeshed with the GE fabric were what sealed the deal.

Ted Smith, the Chief Innovation Officer in Louisville, initiated the meetup between GE and local Makers:

"This is about getting the cycle time down for your product innovation."

"[Companies in] the durable goods category are going to have their lunch handed to them by companies that can move more quickly. [To] introduce new kinds of features and functions, very quickly, the only way you can do that is if you change the way you think about fabrication.

"[At the same time], the reason that these companies have to leave to go to these other markets [like San Francisco or Seattle] is because that's where there's a real concentration of expertise in that industry. That's why they go there. We should just take the lesson, like who should we be attracting here and what are those new products? To me that's the soup that we're trying to cook here. Let's take advantage of the bench strength that's in the industries that we have and let's guide future entrepreneur inventors down the path."

For Smith, that means leveraging both the resources of corporations and the innovativeness of entrepreneurs.

"From an economic development perspective, Making and more modern approaches to manufacturing have a proven and immensely synergistic fit with local industry. As a result, Louisville is now looking to figure out where else this synergy can be applied for their other signature industries. Smith went on to say, "I think there's a lot of room in this movement for this kind of empowerment, but I do think it requires some kind of organizing principle, like FirstBuild is an organizing principle for most scale manufacturing."

Venkat Venkatakrishnan, Director of Research & Development for GE Appliances, explained:

"One of the advantages of having an open business model is we don't have to go through a lot of agreements and bureaucracy to get something done."

Meeting with local Makers at FirstBuild in Louisville

"A number of the universities from this area are engaging with us—the design schools and the engineering schools—and the reason is that this is the best way to produce the best engineers and Makers and artists. We leverage them often.

"For instance, if I want to build 15 ovens tomorrow, we pay engineering students on a part-time basis. They are looking for part-time jobs and we actually get them to come and make the parts and assemble the ovens and learn what good design is in doing so. It's not just an assembly process; it's a big learning process. Now, we get a lot of assistance from the university, and the university looks at it as the best experience that they can give their students in terms of education. And it's not only engineers. We even have artists coming and working in our place and working with us. What we are trying to prove is if you set up an open co-creation facility like this where you not only design products but also build products and sell products, we can do all three... We are now convinced that the future of product development will be similar to the way we are doing things."

FORGING AN ECOSYSTEM TO REFLECT A CITY'S LEGACY AND STORY

Macon, Georgia's SparkMacon

Macon, Georgia has a population of 91,000 people and is located 80 miles northwest of Atlanta. Nadia Osman is Director of SparkMacon, a storefront Makerspace, and organizer of the Make-End Maker festival. Prior to that, Nadia served as Director of Revitalization and Business Initiatives for the College Hill Alliance, a small nonprofit that facilitates the economic development of the area between Mercer University and downtown Macon.

"We wanted Macon to be a place with a new narrative that's attracting and retaining talent and that completely supports Makers. In Macon, they are identifying stakeholders and bringing them together. They are doing a creative census to identify creative industries and the people that work in them as well as their skill sets."

Nadia emphasized the importance of connecting all of the Makers that are involved in the academic, business, government, and nonprofit worlds.

"We have the infrastructure, but they're not talking to one another," she said. The goal is for Macon to become a hub for talent, education, and industry."

Macon has a rich history it can draw upon. It is known for music making, being the birthplace of Otis Redding and the Allman Brothers. "That leads all the way to the manufacturing base that we have as well as the Makers of today who are based in the tech and creative sides," said Nadia. "It's been around a long time. We just haven't had a word for it. Also, we haven't yet successfully capitalized on it.

"We have done the research into entrepreneurs and how they're being supported in the area. We wanted to find out why we're honestly not attracting and retaining talent. What it came down to again was not that we didn't have the infrastructure. We have all of that but we had not found a way to actually bring it together.

We didn't have a way to reinforce the value of taking risks on new ventures, that you can do things that are nontraditional. Our issue right now is fundamentally changing the culture of what it means to operate, to live, to work, and to play in Macon.

MAKERS SELF-IDENTIFY AS "MAKERS" NOT AS ENTREPRENEURS

"When we first explored supporting and understanding the entrepreneurs in our area, that word in and of itself–'entrepreneur'– came across as somewhat pretentious. Existing entrepreneurs did not identify with it. The minute that we put out a call for Makers, even though people did not understand the word, [most] actually gravitated towards it... [they] understood that the Maker is the first word, and it's the first movement that has tied together such different industries. Maker brings together our engineers, our artists, our manufacturers, and our tinkerers. Entrepreneur doesn't do that. No other word does that.

"What has made a lot of this new work in community and economic development possible is that, as most cities have found, if we don't change things, we're going to die. We need to do it now."

University student working on a component for FirstBuild.

MANUFACTURING SECTOR AS A LEADER

Eric Gertler: New York City Economic Development Corporation

The manufacturing industry in New York City, which had been in decline for 50 years, began to stabilize in 2011. Now there is growth in manufacturing. The industrial sector represents about half a million jobs in New York City. Manufacturing is about 15% of that so it's about 75,000 jobs. New York City's Economic Development Corporation (NYEDC) has been focusing on advanced manufacturing and connecting the Maker community in the city. Eric Gertler was formerly Executive Vice President for the NYEDC; he has since moved on to a job in the private sector at Ulysses Ventures. Under his tenure, NYEDC developed a wide range of initiatives, perhaps more than any city in the country.

"We realized that the innovative advanced-manufacturing sector was bubbling up. It was quite dispersed around the city and we devoted much of our early efforts to simply trying to connect some of the Makers with some of the facilities and at least provide some cohesion to that movement. But that is just part of the reason why we've seen a lot of this emerging sector develop in New York City. New York City is also ideally suited for this type of sector because we have a great blend of creative folks, artistic folks, a strong and growing technology sector, and a legacy manufacturing industry. All of this combined with the ongoing and vibrant business that goes on in New York helps create a very robust advanced manufacturing sector in New York City.

"We launched the Next Top Maker competition, which is simply one of the many activities that the city undertakes to help support the Maker movement. This competition is really a challenge to innovators, to Makers, to designers, and to creators to originate and come up with innovative products and ideas that have commercial potential.

"But, given the geographical constraints of and zoning within New York City, we are still trying to figure out the real estate for this sector and its physical requirements. What's interesting, however, is that when you look throughout New York City, the advent of new manufacturing has really been a five borough phenomenon. There are companies like littleBits that was created in Manhattan, MakerBot in Brooklyn, or Shapeways in Queens. We are starting to think through Manufacture New York, as well as the involvement of the Brooklyn Navy Yard and the Brooklyn Army Terminal. Incidentally, these two buildings were built by the federal government and then bought by the City for one dollar about 40 years ago.

"We are starting to rethink how we use this space. There are millions of square feet that could be modified for legacy industries that need new types of innovations. We're not there yet. But we have certainly made great strides. What's going on at the Brooklyn Navy Yard is really interesting. There's some great companies that are doing different types of manufacturing as well as advanced manufacturing in the Brooklyn Army Terminal but I still think we are in the early parts of that movement.

"As a City, we have spent a lot of time focusing on what I would characterize as public-private partnerships. But it's even more than that. It's the combination of business with academia in the City that will have a profound impact on our future. For example, the Applied Science Initiative is a partnership with Cornell Tech to bring and increase the number of applied science students and engineers in New York City. The reason why such partnerships are important is that this new Maker movement has come from all directions. It started from the community, bottom up. It came from business. It came from the academic centers. And, finally, it has developed because the City has been a convener of all these groups, bringing together and connecting the universities with the businesses, with financing, and enabling those connections to happen faster.

"One of the many things we do at the New York City Economic Development Corporation is welcome a lot of groups from all around the City, and, quite frankly, from all around the world. We are excited to share what's going on. But, for now, we are continuing to think through the connections and hone what we are doing things to spur innovation and activity. I truly believe that, with government, little things can lead to big actions. And, this is certainly a sector where you can easily imagine the little things we are doing now leading to another vibrant and growing new sector in New York City."

ManufactureNY and the Reinvention of New York's Fashion District

Bob Bland is Founder and CEO of ManufactureNY, a partnership between the Economic Development Corporation of New York City, private real estate developers, and the fashion industry. Its goal is not just to preserve the fashion industry in NY but also to help it thrive in a new world where fashion called wearables uses electronics and becomes interactive. In its 150,000+ square foot space,

ManufactureNY is offering long-term leases at favorable terms to retain and renew many of the small businesses that make up the fashion industry in New York.

"I've always said that America's been defiant. Our actual reputation around the globe is based on innovation and entrepreneurship, and that's what we need to be doing here now to create a new type of ecosystem not just for creatives, but all the different portions of the supply chain in my community. Every single person who works in New York City has a family. And if they're working in New York City, that's benefitting another business. It amplifies that benefit to do as much as you can here instead of taking money elsewhere.

"The businesses we work with need support over the long term, and it starts with general operations. Overhead is where people are getting killed right now, especially in New York City. Their overhead expenses include their rent, utilities, payroll, all of that. And so any change in their circumstances can be incredibly volatile, especially for contract manufacturers. They're typically working with 10–15 percent margins because that's the price that the customer expects. And there's no way to get out of that model without ballooning your margins and losing all of your customers.

"…[I]f we're not getting our heads together and all coming together in one space to create solutions, then what's going to happen? So I feel that with ManufactureNY, this model of economic development gives people who would otherwise never meet each other a chance to work together and create the fashion supply chain of the future.

"By controlling your entire supply chain and having accountability for everything from the concept to the distribution, you can play with all the levers. And one lever is certainly that direct sales is now roaring back with a vengeance, especially for startup companies in the design space. That can allow you to increase your margins by an enormous amount and then pay a little bit more for the clothing while still maintaining a fiscally healthy business.

"We need to create meaningful solutions and not just rely on the past. So when people come to us and say, 'Oh well, that's impossible,' it's like believing is seeing. We're going to create that here in that way where… with clear vision, with fearless execution, and

Makers self-identify as Makers versus as entrepreneurs

with the right people around to share this vision, we can absolutely make a new economy that supports everyone, not just the people at the top.

"We will bring wearable technologists and traditional fashion designers and contract manufacturers together because we see a lot of promise in these collaborations. And even here in our temporary space, this space which was 150,000+ square feet that we're currently deploying, we've already seen so much of that collaboration resulting in successful ecosystems for each line. So I think it'll be a game changer."

CONNECTING THE LOCAL SUPPLY CHAIN

Manufacturing in Los Angeles

Krisztina "Z" Holly is Founder and Chief Instigator of MAKE IT IN LA, an organization run out of Mayor Eric Garcetti's office as part of its Entrepreneur in Residence program.

Today, Krisztina works in downtown Los Angeles at the La Kretz Innovation Campus, a space that houses the LA Cleantech Incubator, The Prototyping Center, LADWP laboratories, and a training center. She also records her podcast, The Art of Manufacturing, at Maker City LA, a workspace collective for creative businesses, Makers, and entrepreneurs, housed on the 11th floor of The Reef in downtown Los Angeles in what used to be the LA Mart, a huge space for wholesale showrooms.

Holly explains how Los Angeles is connecting its vast manufacturing resources:

"Los Angeles is the largest manufacturing center in the country. Manufacturing should be as influential and dynamic as tech and Hollywood. These two industries are so big and important that they're spectator sports now, for better or for worse. Manufacturing is not that. It's underground and people don't think it's sexy. Manufacturing hasn't really gotten the respect that it deserves, especially considering we literally have four times as

Brooklyn Navy Yard as viewed from the air

many jobs in manufacturing as we do in film and television. You don't hear about that, you hear about Hollywood.

"We did a year-long study of local businesses last year and learned some surprising things. If you look at the Dunn and Bradstreet data, there are about 30,000 self-identified manufacturing companies in LA County. That's a lot. They span industries from aerospace and electronics to food and fashion. Most of them are very small and some of them are mom and pop shops. Others are highly capitalized high tech enterprises, like SpaceX and Hyperloop One.

"The folks that we've engaged with are incredibly diverse, and while this is not 100% representative of the industry, our respondents are nearly half women- or minority-owned. Pretty cool: 47 percent of the businesses were either women-owned or minority-owned. I think that there's a real opportunity to also recast the image of entrepreneurship, and of manufacturing, through the people that are rolling up their sleeves and taking risks on their next big thing.

"We found that 58 percent of the businesses that we surveyed have excess production capacity, and they're interested in learning more about contract manufacturing. Meanwhile, most companies in L.A. County are interested in sourcing more locally but they don't really know how to connect to the right capabilities at the right price.

"There's a lot of recent activity and startups trying to create tech platforms for connecting like BriteHub, ThinkFab, Connectory, Maker's Row, and more. The feedback that we heard from manufacturers was that there's skepticism around a pure technology solution. They're open to using technology-enabled approaches but they really want to build relationships face-to-face. They value recommendations from their peers and their colleagues.

"And in certain industries, like apparel, there's a lot of secrecy around who your suppliers are because, for this old guard, that's their secret sauce. There's value in having those networks. Companies are less open to sharing their networks. Now, there is a subset of these companies that are really interested in

ManufactureNY is reinventing the fashion district.

collaborating. So the key is not quantity; at this point, we need to focus on engagement. We're identifying the innovative suppliers and customers in this ecosystem that want to work together to grow the pie.

"We've started developing a coalition of organizations whose mission is to support the entrepreneurial ecosystem and inspire Makers to turn their ideas into products in LA. As a group we're committing to developing educational programming to serve more than 1,000 entrepreneurs in the coming 18 months."

GIGABIT BROADBAND AS AN ENABLER

Inside ManufactureNY

"The Gig" Comes to Chattanooga, Tennessee

An interesting example of city investing in network infrastructure is Chattanooga's city-provided ultra high-speed fiber-optic internet service it calls "The Gig."

Established five years before Google prototyped fiber in Kansas City, the network transfers data at one gigabit per second: 50 times faster than the connection for homes across the rest of the U.S. It's among the fastest internet in the world. Surprisingly, the network was not built by a large tech or telecommunications player like Google or Comcast. The publicly owned electricity company, EPB, built it as a way to monitor and communicate with advanced digital equipment being installed on the grid and respond more quickly to outages. While the fiber itself is impressive, the doors it is opening for the city are much more exciting.

Mike Bradshaw is Founder and CEO of Company Lab (Co.Lab), a nonprofit entrepreneur center in Chattanooga. He sees the potential in gigabit broadband to support advanced manufacturing by allowing very complex objects to be designed, sourced, and built in concert with suppliers, manufacturers, and experts in advanced materials who are geographically dispersed. Zero latency becomes a requirement if the dispersed team of players involved in advanced manufacturing are to collaborate together effectively.

"What would happen if these machines were hooked up, in a fluid network where geography and time weren't really an issue, meaning that you'd have photonic networks with virtually zero latency between the machines, allowing them to take a complex build and adjudicate the manufacturing process in this set of geographically distributed machines?"

This network dissolves barriers to information transfer, empowering local entrepreneurs and Makers to collaborate around high value domestic manufacturing without limitations that come with geographical distribution.

Sheldon Grizzle, founder of the Company Lab, speaking to The New York Times, had this to say about "The Gig," the taxpayer-owned, fiber-optic network:

"It created a catalytic moment here... [it] allowed us to attract capital and talent into this community that never would have been here otherwise."

Its impact on Chattanooga is only the beginning. The intention is to advance the technology even further to promote collaboration as an engine of innovation and economic development where entrepreneurs, innovators, and Makers can connect in real time with cities across the nation, putting Chattanooga and every city that joins them on the innovation map.

IMPLICATIONS FOR CITIES

The most notable examples of organizing structures are those that leverage the formality and resources of partnerships to augment the openness of knowledge sharing and the development of talent through networks.

Encourage the Maker City ecosystem to grow from the edge to the center. Organizations that are explicitly "about" Making are a great starting place as your city builds out its ecosystem. But the most robust ecosystems we encountered were open; they allowed anyone with both a strong interest in Making and shared values to participate. Faith-based organizations, trade unions, and members of the scientific and medical communities may seem like unlikely members at first, but their contributions can accelerate change at the center of the Maker City ecosystem in your city.

Map what you have. Encourage entities in your city to place themselves on the map using open-directory tools. We talked about two such tools in this chapter: themakermap.com and 3dhubs.com

Survey people and organizations who are involved in Making in your community, to learn what they need and where they fit into the ecosystem. A sample survey is available on the last two pages of the Portland Made report to use as a starting point. The methodology is pretty straight forward. Portland Made worked with a research team through a university, which may be a no-cost or low-cost way for your city to get this kind of survey completed.

Analyze your ecosystem and look for gaps. One cost-effective way to do this is to hire a college student to work as an intern one summer and map the resources that exist using the spider diagram introduced above which distinguishes between resources on the edge of your ecosystem and resources at the center. Once your city has identified a major gap in its ecosystem, it knows where to focus its economic development resources.

 Start organically. The most robust ecosystems start out organically in a self organized fashion; only later, as the ecosystem matures, does it make sense to ask an organization to step in and coordinate activity. To organize Maker activity in your city, often you need look no further than the people who run the local Makerspace. But there are other organizations that can help. Many cities are finding they benefit from setting up a group that encourages pride in locally made goods and/or providing explicit support for urban manufacturing.

Consider municipal broadband. Chattanooga and other cities are finding that the introduction of gigabit broadband as a municipal utility, much like electricity, water, or waste disposal, can have a transformative effect, particularly on advanced manufacturing in a Maker City.

Look at real estate as an asset. Brooklyn has a strong Maker ecosystem in no small part because it has industrial space where Makers can thrive at the Brooklyn Navy Yard and in a facility for the fashion industry at ManufactureNY.

Study the economic impact of Making in your city.
We believe that Making is having a transformative effect on our cities, creating jobs and economic opportunity. But we cannot always prove it. A cost-effective way to get started here may be to identify an economist at your local college or university with a research interest in the Maker movement.

Hire or appoint someone to serve as a dedicated advisor for Maker activity in your city. San Francisco experimented with this and quickly discovered that what they needed was less of a Maker liaison and more of a champion for Making that led directly to urban manufacturing. They provided us with a sample job description for a "Make to Manufacture Fellowship" that cities can leverage.

RESOURCE   QUICK
                  WIN

IMPLEMENTATION   POLICY   BIG
ADVICE   IDEA

EDUCATION

LEARNING IN THE MAKER CITY: A QUIET REVOLUTION

For over 50 years, technology has raced ahead of the traditional ways we educate our children, often to national consternation.

Marshall McLuhan (1964) is best known as a media scholar; he was also prescient in anticipating this impact at the dawn of the information era:

"The electronic environment makes for an information level outside the schoolroom that is far higher than the information level inside the school room. In the nineteenth century the knowledge inside the schoolroom was higher than the knowledge outside. Today it is reversed. The child knows that by going to school he is, in a sense, interrupting his education."

McLuhan argued that education needed to shift from

"stenciled instruction to discovery, probing and exploration."

McLuhan went on to say,

"The young today want roles, and total involvement rather than specialized jobs or goals."

McLuhan asked in 1967,

"How can our youth look forward to specialized jobs when the computer world of automation may eliminate those jobs?"

McLuhan spoke for a whole generation of educational experimentalists who saw that project- and discovery-based learning can both powerfully engage students and better prepare them to succeed in life.

McLuhan follows in the footsteps of **John Dewey**[1] who argued for social interaction and experience-based learning in the early twentieth century, **Jean Piaget**[2] who argued against passive educational models, and **Seymour Papert**[3] whose educational constructs developed in the 1980s set the standard for how to use technology in the classroom to enhance discovery-based education.

The Maker movement takes this discussion to the next level by turning the city into a learning community, one where parents, educators, and students come together to reshape learning around the needs of learners. The learning community inside the Maker City is leading the way in asking questions that will define education for the next 50 years: How do we make education more engaging, more relevant, and more collaborative? How do we break down the silos between disciplines and teach our students in a more integrated way? How can we be sure that what we are teaching our youth is aligned with the opportunity for meaningful work we expect to be available to them upon graduation and beyond?

EQUITY IS AN ISSUE

As we'll see in this chapter, there is a remarkable range of Maker-based education experiments that are working today to improve outcomes, excite children, and reduce dropout rates. The issue is equity: how to bring these advances to a wider audience. One of the goals of this book is to demonstrate to school districts, policy makers, parents, and funders that there is proven value in exposing kids to Making at an early age, so they can use the tools of Making to solve problems, to gain confidence, and to develop a sense of agency.

Formal schools are not the only or even primary place where learning happens. The learning community inside a city spans not just schools ("formal learning") but also a set of loosely coupled network of informal learning opportunities that exist inside the Maker City. The informal learning environment can include: libraries, museums, camps, after-school programs, community centers, rec centers, churches, and universities.

Many out-of-school programs offer exceptional learning opportunities, but equitable access remains a staggering barrier. Economists Greg Duncan and Richard Murnane (Source: **New York Times 2012**[4]) have found that enrichment spending by affluent families was nearly 700 percent greater than that of the poorest families.

To truly remake opportunities available to our children, we need more collaboration between individuals and organizations across the entirety of the learning community inside the Maker City, not just schools. Financial support is also a "must," which means extending the reach of the learning community to embrace foundations, nonprofits, and corporations for financial support.

How the Maker Movement is Changing Formal Learning

Inside schools, change happens slowly because education is a system that tends to be entrenched and highly resistant to change.

This is why Gregg Behr, whom we met in the Ecosystem Chapter, cautioned the team interested in "remaking" learning in Pittsburgh to start small, with one classroom, one school, and one district at a time.

ELIZABETH FORWARD SCHOOL DISTRICT

The Maker movement has been transformative for the Elizabeth Forward District, a relatively poor district located 15 miles from the center of Pittsburgh, Pennsylvania, just on the edge of Appalachia. This was a district with both a dropout problem and opt-out problem, with over 70 students opting for charter schools or online schooling. Student test scores were middle of the road, and the school struggled to engage its students.

The district's Superintendent Bart Rocco and Assistant Superintendent Todd Keruskin turned to Pittsburgh's research universities, education technology companies, cultural institutions, philanthropies, and civic leaders to transform their district from "frontierland" to "futureland."

The Aha! Moment
The change started by partnering with a major university, in this case Carnegie Mellon University (CMU). Rocco and Keruskin met with Don Marinelli, Co-director and Founder of the Entertainment Technology Center (ETC) at Carnegie Mellon University, who asked:

"Why don't your schools look more like theme parks? And why aren't you thinking really differently about integrating left and right brain thinking like we do at Carnegie Mellon and the ETC?"

Bart Rocco went down to the Entertainment Technology Center at CMU and that was his "aha" moment. He saw what was happening there, he saw what they were doing, and he said," Oh my God, we are not preparing our kids for this future. And this isn't even the future, this is happening right now."

As a follow up to this meeting, Bart visited **Quest to Learn** School in New York and the **YOUmedia** space, a set of youth-oriented Makerspaces located in branches of the Chicago Public Library.

The Importance of Seed Money

The Grable Foundation provided the Elizabeth Forward District with seed money in the form of a $10K grant. That $10K grant enabled them to change just one classroom in their high school building to mimic the Entertainment Technology Center at Carnegie Mellon. Called the Entertainment Technology Academy, the initial focus in this one classroom was on game design, leveraging instructional kits provided by Zulama, a company had which spun out of the master's program at ETC.

Over 25 percent of the students at the high school ended up wanting to enroll in courses offered through the Entertainment Technology Academy.

The Dream Factor at the Elizabeth Forward Middle School

Bart Rocco's "Aha! moment" helped spark the genesis of the Dream Factory at the Elizabeth Forward Middle School.

The Dream Factory is a set of studios that are linked together and designed to enable students to pursue projects that are of interest to them, at their own pace, and in their own way. Teachers provide exercises that help cement selective learning topics at a pace that makes sense for the individual student.

Originally, the studios started out in separate rooms. Very quickly, teachers and administrators decided to make the space work

QUEST TO LEARN
q2L.org

YOUMEDIA
mcbook.me/2cqx2pp

more like the YouMedia space in Chicago, knocking down the walls among and between computer science, industrial technology, and the arts to create one large Makerspace.

According to Gregg Behr of the Grable Foundation, the Elizabeth Forward Middle School was the first public school in the nation to dedicate space to a Makerspace/fablab and to integrate project-based learning around the tools of Making into their instructional practice.

Today, Elizabeth Forward Middle School offers computer programming to all 6th graders and robotics courses to all 7th graders. After the students learn the skill sets in 6th and 7th grade, 8th grade they can make their dreams come true by designing and building almost anything, taking advantage of the of the Dream Factory's equipment, which includes both software (2D and 3D modeling, animation) and hardware (laser cutter, CNC router, 3D printers).

The Dream Factory is more than a Makerspace embedded within the school. Students are supported by a curriculum at Elizabeth Forward Middle Schools which ignores the formal distinction between disciplines in favor of a more integrated approach.

What we see happening here is that the Maker movement is breaking down the barriers that can be in the way of real learning for students, enabling them to have a sense of agency around what they learn and to explore their passions and dreams.

Results

In just the past few years, the Elizabeth Forward School District has seen dropout rates fall drastically, from 24 students in 2009 to one in 2012. Enrollment in the Entertainment Technology Academy has exploded: from a mere 29 students in January of 2012 to over 200 students in multiple courses during the 2012–13 school year. Out of the 497 school districts in the state, Elizabeth Forward has leapt from #240 in 2009 to #155 this year, moving up 82 spots in four short years.

NUVU INNOVATION

Another experiment in education and the Maker movement can be seen at the NuVu Innovation School, a full-time innovation studio for middle and high school students based in Cambridge, Massachusetts, just a stone's throw away from MIT.

NuVu designed its curriculum to better meet the needs of students who do not learn in traditional ways and often feel like failures because of this. NuVu prides itself on giving its students a sense of agency, putting the student in charge of what they learn and how they will apply that learning to real-world problems.

To do this, the team at NuVu eliminated the standard stuff of middle and high school and rebuilt almost the entirety of its curriculum around Makerspaces and a studio-based approach that Seymour Papert (2001) and others call **project-based learning**[5].

At NuVu there are no courses, subjects, one-hour schedules, or grades. Instead, the focus is on teaching students how to solve real-world problems, from inception to completion, by prototyping and testing. Students can either enroll for a single semester or trimester or for the entirety of their middle school and/or high school education.

For example, after learning the basic tools of Making, students might work in small teams to create an innovative solution to a problem like this one:

People who are handicapped and live in cities may live in small apartments where wheelchairs don't maneuver all too well. If you have to get out of the wheelchair and get to the toilet, someone has to lift you, greatly inhibiting any sense of independent living.

In response to this challenge, students at NuVu designed a device called "Uplift" which helps a wheelchair bound person lift themselves up.

When you visit NuVu you are hit with undeniable energy, a sense that this is a place where young people are excited to come everyday, to get coached in the tools of Making, and to learn how best to apply those tools to solve problems. Importantly, at NuVu young people get the guidance they need to follow their passions and to work on problems that interest and excite them.

One young woman came to NuVu with low-self esteem and very poor grades; both parents and teachers at her "old school" believed she was headed for delinquency or to drop out of school prematurely. Today, she attends MIT and is preparing herself for a career in robotics.

TOWARDS A NEW PEDAGOGY THAT GRADUATES ALL STUDENTS INNOVATION READY

Tony Wagner is Expert In Residence at Harvard University's new Innovation Lab and a Senior Research Fellow at the Learning Policy Institute. He writes and speaks about the importance of graduating every student as an innovator through a "combination of play, passion, and purpose... to develop the discipline and perseverance required to be a successful innovator."

Wagner calls for formal learning institutions to rethink the pedagogy they rely upon in classroom settings in five distinct ways:

1. **Create a classroom culture not of individual achievement, but instead of collaboration.** It is remarkable the extent to which we find again and again that innovation is deeply rooted in the power of collaboration, and we must resist the myth that inventors are creative by being solitary, uniquely creative, individual geniuses.

2. **Focus not on specialization but on teaching that cuts across multiple disciples.** Curricula should be wide-ranging, covering many topics and subjects, without regard for narrow disciplines or boundaries.

3. **Practice not risk avoidance but experimentation.**
   Enable students to iterate and to explore through trial
   and error. It's important that students learn they can try
   something, fail, and then try something different. "At the
   core and essence of innovation is the ability to learn from
   failure which means facilitating an environment where
   students have to take risks.

4. **Promote the value of creating, not consuming.** Make
   classrooms places where students work and are serious
   about producing genuine, concrete, meaningful products.

5. **Encourage children to motivate themselves
   by underlining intrinsic rather than extrinsic
   incentives.** Encourage exploratory play, the finding and
   pursuit of a passion, and the idea of giving back.

Wagner's thinking here strongly aligns with changes in formal
learning wrought by practitioners as they bring the tools of Making,
the Maker mindset, and Maker ethos into classroom settings.

To Wagner's thinking we would add one additional element,
borrowed from NuVu: the importance of portfolio-based
assessment. A portfolio serves as a compilation of a student's
work and is meant to show the student's growth over time and
development of key academic and life skills (creativity, critical
thinking, collaboration, communication, research, quantitative
reasoning, and analysis).

MIT was one of the first universities in the country to enable
students to submit a **Maker portfolio**[6] as part of its application
process. (Portfolios are important not just for college admissions
but also to present to potential employers, as we'll see in Chapter
5 on Workforce Development.)

As universities like MIT and Yale embrace the Maker movement, it
validates the efforts of the K-12 educators, who are champions for
Making inside their schools.

The **MAKER PORTFOLIO**
is an opportunity for
students to showcase
their projects that require
creative insight, technical
skill, and a 'hands-on'
approach to learning by
doing. Members of the
MIT Engineering Advisory
Board review all Maker
portfolios. If you would
like your technically
creative work to be
reviewed by academic
and instructional staff,
then it might be a good
fit for the Maker Portfolio.

UNIVERSITIES: OPENING UP ACCESS TO THE TOOLS OF MAKING

Increasingly, universities are making their Makerspaces available to all students, not just those in a particular class or a particular major and even, sometimes, to the entire community, as we'll see below.

Understand that this isn't something that the institutions want to do. It's something they have to do, to remain competitive.

"Institutions simply are going to have to have Maker experiences available in multiple forms, and across multiple programs," said Timothy McNulty at Carnegie Mellon University. "That will increasingly be a defining feature of the holistic nature of the institution, and therefore its competitiveness for ambitious and bright students: to respond to students' desires to be efficacious, to have agency, and to recognize the role that Maker activities play in those overall ambitions." This change is redefining the nature of the flow and cycle of campus activity and the points of intersection between different campus groups. McNulty has seen a waterfall effect in the rethinking of policies, curricula, programs, and approaches to community engagement.

Yale
Yale turned their Engineering Library into a Makerspace available to any student, regardless of their field of study. Hundreds of students became members in just the first three months, taking safety training to be able use the machines in the space. One of the faculty champions for the space said the university was surprised by the pent-up demand. "We had no idea how many student projects were happening in dorm rooms," he said.

Georgia Tech University
At Georgia Tech University, the Invention Studio was opened in the Mechanical Engineering Department as a space for engineering students to develop capstone projects as seniors. Now it is open as a Makerspace to anyone on campus. It has also become a social space where students hang out between classes. It connects

engineering students with students from other disciplines and creates the possibility of collaboration.

Arizona State

At Arizona State University's Chandler campus, there is a Makerspace run by TechShop that offers free access to students and paid access to community members.

Case Western Reserve University: Think[box]

Think[box] is a seven story, 50,000 square foot Makerspace, built at a cost of $37M at Case Western Reserve University in Cleveland, Ohio. Think[box] started small and operated for several years on a mostly volunteer basis before getting the attention of the university administration. The new version opened in the fall of 2015. Unlike many university Makerspaces, it is open to the entire community, not just students, for free.

According to Lisa Camp, Dean ovf Strategic Initiatives for Case Western Reserve University:

"It's really an economic development engine in many ways."

Each floor represents a specific programmed activity within the whole building.

It starts with the first floor, which serves as a social gathering space where community members bump into one another.

The second floor is where community members and folks on campus from different disciplines off campus can collaborate and come up with new ideas.

Think[box] at Case
Western Reserve
University

The next floor is devoted to prototyping using 3D printers and laser cutters, traditional Makerspace equipment.

The fourth floor is more like a traditional fabrication shop in factories where you have woodworking equipment, welding tools, and other machine shop equipment. The next floor is dedicated to entrepreneurship where you can get legal advice and even funding. Finally, the top floor is incubation space for small startups to work in.

"It's not just an engineering project; it's meant to be accessible for all disciplines on campus," said Lisa. "It's also open to the community because we believe in this kind of openness. People bumping into others from different walks of life and bringing different areas of strength is a good thing."

Together, what is happening in K-12 and Higher Education around Making amounts to a cultural shift. Instead of focusing on imparting content to students, educators are focused on enabling students to learn more collaboratively and to focus on solving real world problems of interest to them.

BETTER ALIGNING HIGHER EDUCATION WITH REQUIRED JOB SKILLS

Luke Dubois is Co-Director of the Brooklyn Experimental Media Center at the NYU Tandon School of Engineering (formerly known as NYU Polytechnic Institute), the largest private university and the second oldest engineering school in the United States. While he teaches engineering, Luke is also a musician and an artist. He looks for cross-disciplinary interests in his students.

Third Floor at Think[box]

Everything here is an 'and': engineering and soft circuits, fashion and sensors, art and advanced manufacturing.

This diverse student body from all five boroughs has traditionally seen engineering as an aspirational entry point to white-collar work.

"Historically you didn't want to get your hands dirty because you're going to engineering school to be the guy who looks at the blueprints and orders around the people who get their hands dirty."

But in the last few years there's been a shift:

"We're finding now is there's this really interesting strategic inversion in that thought process where the more you get your hands dirty the more innovative you are."

"With our students, we're doing all these classes now that are all about project-based learning. You're just constantly Making. It's almost like art school, it actually feels to me like art school or like conservatories. Our students are just building and prototyping as a matter of course. Not even for a class, it's not even that they're doing their homework, they're just tinkering. And that kind of tinkering thing creates whole new avenues of expertise, like you get your chops up in different ways. It opens up different avenues to jobs."

What worries Dubois the most is making sure to keep the university aligned with required job skills, skills that cannot be easily anticipated due to the fast pace of technology change.

"The real pain point to me is where if education can't keep up with all these changes in industry. If people are still educated to be truck drivers and the truck driver jobs go away, then that's an enormous loss. I think what really needs to happen is just everybody needs to keep their eye on the ball as these industries transform to say, 'What are the requisite educational requirements?'"

We'll talk about this issue of alignment in the next Chapter on Workforce Development.

THE ROLE OF THE EDUCATOR: FACILITATOR, MENTOR, COACH

In a Maker City, educators are not less important. They're more important. The only thing that has changed is their role: from a lecturer who talks to students in a highly prescribed way, to a mentor, coach, and facilitator who is responsible for guiding the student into a mode of learning that is hands on, encourages students to take on real world problems, invites discovery and rapid prototyping of what works, and supports collaboration with others as the fundamental way problems get solved.

In short, the Maker movement is changing education from the inside, one educator, one classroom, and one school system at a time. It's a revolution but one that in many ways is being kept

on the "down low" to avoid the inevitable shouting match that happens in America every time we talk about pedagogy and what and how we teach our children.

There is pent up demand for change. We've already talked in Chapter 3 on Ecosystems about the Remake Learning Network and how it grew from a simple pancake breakfast with 10 people to encompass 250 institutions and 1,000 educators in the greater Pittsburgh area.

We see a similar phenomenon happening nationally, thanks to an initiative called **The Maker Promise.**

This is a joint initiative of two nonprofits: **Digital Promise** and **Maker Ed.**

The Maker Promise launched in March 2016, coincident with the White House as it announced the National Week of Making for 2016.

In just a few months, representatives from 1,200 K-12 schools across 49 states signed the Maker Promise thereby committing to take action on Making in their schools and communities.

What's striking about the Maker Promise is how simple it is. It asks school leaders to do three things to encourage Making in the schools.

→ Create a dedicated space for Making

→ Designate a champion of Making

→ Display what your students Make

President Obama interacting with young Makers

MAKER PROMISE
makerpromise.org

DIGITAL PROMISE
digitalpromise.org

MAKER ED
makered.org

Where the Change Started: With Informal Learning

The change that is happening in the learning community started with changes in informal learning inside our cities. Every city has places where informal learning happens, in the form of libraries, museums, parks, and community centers.

LIBRARIES

Before anyone used the term "sharing economy," libraries were a place where the tools of education were available to share. Starting in the 1890s, the tools that mattered were books to build a more literate workforce. In the 1990s, libraries started to move away from the physical book to focus on providing access to the internet and to multimedia tools. Today, libraries are embracing Makerspaces, in their role of providing a centralized place where learners can share the new tools of learning and production.

Libraries are very much at the edge of the Maker movement in education meaning that the change that happens there has a kind of ripple effect on change happening elsewhere in the learning community, both formal and informal.

Clevelend Public Library: Cleveland Ohio

The Cleveland Public Library calls itself "The People's University."

We talked with CJ Lynce, the Makerspace manager at the Cleveland Public Library. In the beginning, CJ tells us, the most popular materials in use within the library's Makerspace was duct tape to create flowers and wallets. Today, the most popular items are **littleBits**, a kind of electronic kit with building blocks that are color coded and snap together with magnets, enabling students to invent new electronic devices with very little effort.

"[LittleBits] were the absolute least used piece of equipment or tool that we had in our arsenal. They sat on the shelves for two years. We just recently brought them back out, and they have actually become the most requested kits that we have. So much so that we just purchased a couple of the Pro kits, and we are going to start sending them out to branches because it's given

LITTLEBITS
littlebits.cc

them the opportunity to experiment with this idea. This is very reflective of the transition that we've seen in the past year but I'd say especially in the past six months."

San Diego: Central Public Library

Misty Jones is Director of the San Diego Central Public Library, which started with a small single-room Makerspace. "We just put the walls up and put the furniture in for a new Makerspace on the 3rd floor that's about four times the size," said Misty.

Just buying equipment for a Makerspace is not enough.

"We hear that from a lot of libraries, calling us to say that they put 3D printers in and nobody uses them," said Misty. "It is about the culture, and getting people involved. And that's what my staff has been so great at, getting out there in the community and being part of this Maker culture. And that's what's made us so successful." The library has also been successful recruiting volunteers for the Makerspace.

"We had a paraplegic that used to ride bicycles," said Misty. "He has a hand crank and he could never get a bicycle that he was able to crank with his hands. So he came into the lab and started using it and designed his own. Now he's working on getting it patented."

This new Makerspace will be called the "Innovation Incubator."

The Innovation Incubator at San Diego Public Library

The library has seen the number of patents filed increase significantly since opening its one-room Makerspace.

The library also has a focus on education offering STEM and STEAM classes. They hosted the 24-hour Code Day and had about 150 middle and high-school students spend the night in the library coding. The high-school students who come to Central Library are low income; 70 percent qualify for the free or reduced-price lunch program at their schools. "They can come here and be exposed to technology," said Misty. "What we're trying to do is even the playing field so that we can get everybody involved."

MUSEUMS

Museums too are getting in the act, establishing Makerspaces that encourage learners both young and old to work together on projects.

Pittsburgh Children's Museum

Jane Werner, Executive Director of the Pittsburgh Children's Museum, believes that the value of Makerspaces goes beyond STEAM, engineering and math. Their first Makerspace was built for about $5K. "We found when we did this little prototype that we had people staying for 30 minutes to two hours at a time, making things. And I often tell the story that I walked in and there was a mom who was sewing with her two boys. They were sewing a whale and a sand shark. They had looked up images on an iPad. They actually designed them, they were sewing them, and they were stuffing them. And I said 'Wow, those are really nice.' She said, 'Yeah. We've been here so long that we haven't even had lunch,' and this is like three o'clock in the afternoon." She has other examples of a grandfather sewing a pincushion for his wife.

"I sew," said Jane. "I often say the reason I am the executive director of anything is because I sewed when I was a kid, and it gave me the confidence to know that I could make a difference and I could change things, I could change my world by sewing. I could make things go from two dimensions to three dimensions. That's a game-changer when you're a 10-year-old."

Jane built out MAKESHOP in collaboration with Carnegie Mellon University (CMU), which helped them integrate technology with the DIY ethic of the Makerspace. The goal was connecting physical and digital experiences. She appreciates the learning value of producing something. Yet she sees more that happens in the museum context. Jane adds: "Making promotes conversations between the facilitator and the visitor, between the visitors, between parents and grandparents and kids. I'm really beginning to believe that that's the role of museums anyway, to start conversations. And MAKESHOP is one of those places where everybody just feels comfortable making things."

The museum offers professional development programs for teachers, exposing them to the practices of MAKESHOP. They are working outside of Pittsburgh, creating a network in West Virginia. "We're actually going into West Virginia and putting some of our teaching artists down there for one day a week for the entire school year, in one classroom and one school so that they get totally integrated and can do professional development with the teachers right on site," Jane explained.

MAKESHOP at the Pittsburgh Children's Museum

The work of Pittsburgh Children's Museum in collaboration with CMU underlines the role a learning community can play not just inside a city but throughout the region: the professional development program enables schools in 25 sites in West Virginia to create Maker programs that are tailored to the needs of their sites and students. The goal is to create independence, not dependence. "We're hoping to create this network so that we can leave. Then West Virginia will be on their own, and not dependent on us," said Jane.

COMMUNITY CENTERS

Finally, we see community centers, churches, and other places where people congregate in their free time establishing Maker spaces.

Detroit: Mt. Elliott Makerspace

In Detroit, Jeff Sturges created the Mt. Elliott Makerspace at the Church of the Messiah. Jeff moved to Detroit from NYC and wanted to help develop a Maker culture in Detroit. He brought with him past experience at MIT FabLab in the South Bronx and NYC Resistor Hackerspace in Brooklyn. He started a Makerspace in

Mt. Elliott Makerspace in Detroit

Eastern Market for people his own age but sought to develop a community space that could make a difference in Detroit's struggling neighborhoods.

"I really focused on a neighborhood," said Jeff. "I wasn't trying to do something en masse, per se. I was focusing on one neighborhood as a pilot. I didn't have the specific intention of creating a specific business or doing job training." He created a Makerspace in the church basement and ran Maker programs for children on Sunday mornings. The church already had a community garden and a bike shop so introducing soldering and electronics workshops seemed a good fit. He got a grant for about $100K a year through the Kresge Foundation to run the program.

"What was magical about the church was that it was multi-generational. It was all ages. There was a lot of peer teaching that went on, peer learning, and that was a specific intention that we set for people. If you learned something, you got to teach somebody... So the culture that we built around Making was amazing and very strong."

"One of the things we did well is that we had multiple offerings for many people. You could make a T-shirt. You could fix your bike. You could learn electronics. You could play music. You could make music. You could do something in the wood shop," said Jeff. He also realized that many of the people he served needed to see how this might help them earn a living. "A lot of people were like, 'That's cool, but how's this gonna make me money?' We started to develop connections between what we were doing and industry," he said.

One way to look at this activity in an informal learning community built up around the learner is as an equalizer. Thanks to these efforts, students don't have to go to a private school like NuVu Innovation or wait for their school district to reconfigure the K-12 experience around Making. They can go hands on with the tools of Making right now, at their local library, museum, or rec center. Students and parents then bring the excitement back to educators. The movement spreads through the efforts of the Remake Learning Network and the Maker Promise to catalyze additional changes inside school districts.

LRNG: CITY AS A CLASSROOM

What happens when you connect up young learners with all the resources a city has to offer, so that they can use the tableau of the city as a kind of virtual classroom?

This is the intriguing question that LRNG, an innovative nonprofit organization funded by the MacArthur Foundation, is attempting to answer with the help of a long list of for profit and nonprofit partners.

LRNG comes to this effort having observed that young people do better when what they learn connects up with the real world. To that end, LNRG has developed a set of playlists that describe how students can interact with the city as a classroom to earn badges. Not only do the digital badges verify learning has occurred in formal and informal learning spaces, they unlock opportunities for youth to connect to internship opportunities, additional school credit, priority selection of college classes, and more. Equally as meaningful, participants can use their digital portfolio to showcase work to future employers or in college applications.

Currently there are five types of playlists available:

→ **Interest-driven:** Engage in sequenced learning experiences organized around themes of interest and relevance;

→ **Production-centered:** Create a wide variety of media and content in experimental and active ways;

→ **Socially-supported:** Form relationships with peers and caring adults that are centered on interests, expertise, and future opportunities in areas of interest;

→ **Openly-networked:** Access learning resources–including assessments, badges and certifications-across all learner settings; and

→ **Opportunity-oriented:** Earn badges that unlock real-world opportunities (e.g., field trips, workshops, job mentoring, job shadowing, internships, etc.).

In the summer of 2016, LRNG worked with its partners to co-design national Playlists that connect passion to purpose by providing learning opportunities in career readiness, civic tech, game design, personal finance, design, and more. Current Playlists include: The Business by Gap, Inc.; Know Your Power, Know Your Employer's Power and My Power Plan, all designed in collaboration with D.C. and Chicago workforce leaders; Young Money, designed in collaboration with WE; and Be Payday Ready, designed in collaboration with LRNG Chicago. In addition, local LRNG cities are creating their own Playlists. For example, LRNG West Sacramento has created a Playlist that allows young people to intern at City Hall in various departments.

LRNG Uses Badges as Evidence of Learning

Digital badges are publicly shareable digital credentials that provide evidence of a substantive learning outcome and can unlock opportunities, such as job opportunities, internships, or course credit. What sets digital badges apart from traditional degrees or credentialing systems is that they are a form of assessment which contain metadata that indicate specific claims about what the badges represent and can show examples of learning products that support such claims, allowing for greater specification and transparency about requirements and evidence.

More specifically, badges earned through LRNG:

→   Link to examples among a broader portfolio of work supporting the earned badge, including the evidence produced from the learning experience;

→   Convey learning pathways and associated accomplishments along the way, especially when a layered approach is used to communicate sequences or networks of learning experiences;

→   Become part of the learner's lifelong portfolio and shows concrete evidence of learning accomplishments—like positions held on a resume or examples of youth-produced work in a portfolio; and

→ Unlock real-world opportunities for youth, such as showcases of work, job shadowing, or internships.

On June 27th, 2016, the U.S. Conference of Mayors passed a **resolution**[7] that encourages cities nationwide to embrace digital badges for workforce development, employment, financial aid, and higher education. The resolution also encourages city leaders to leverage the LRNG learning platform as a shared digital badging framework.

To learn more about LRNG platform, please visit about.lrng.org and consider downloading the LRNG Partner Handbook from about.lrng.org/cities-organizations.

• • •

How Does All This Get Funded?

Great question.

CROWDSOURCING: KICKSTARTING MAKING

Yancey Strickler is Co-founder and CEO of Kickstarter, a crowdsourcing site where you can put up information about your project and seek financial backing from the Kickstarter community. Yancey came to Pittsburgh and visited the Pittsburgh Children's Museum and kind of fell in love with it.

He said to the museum's Executive Director Jane, "So what do you want to do?" And she replied, "Well, I want to put a Makerspace in every school in America. And I want to use Kickstarter as the platform to do it." That is, use Kickstarter to raise funds from the community to develop Makerspaces in schools and get professional staff development as well. "We just did a beta test of it with ten very different kinds of schools in the Pittsburgh area—rural, suburban, inner city, private, charter, like, all over the place," said Jane. Seven of the schools were successful in their Kickstarter campaigns." These seven schools raised over $109K.

Jane and her team have developed a Playbook for other schools to participate in this Kickstarting Making Program. They have made the commitment to do 75 more schools and to develop Kickstarting Making as a national program. "We're going to have our friends at libraries and museums across the U.S. be the host sites for professional development workshops for teacher bootcamps," said Jane.

CORPORATE PARTICIPATION

Corporations are funding STEM outreach programs to help improve diversity in these fields as well as increase the pool of talent available for hire.

San Diego: Qualcomm's Thinkabit Labs

Thinkabit Labs, an outreach program developed by Qualcomm in San Diego, serves 6th, 7th, and 8th grade children from local schools. Its goal is to bring students in and introduce them to technology and invention through hands-on learning. It exposes many students for the first-time to careers in STEM fields. Ed Hidalgo, who runs Thinkabit, says "How can a kid aspire to a career they don't know exists?"

Susie Armstrong, Senior Vice President of Engineering at Qualcomm says that Thinkabit Labs "is not just a Makerspace, it's a STEM learning experience." It has served over 3,000 kids from area schools, and engaged hundreds of teachers and administrators. The program grew out of the diversity teams at Qualcomm trying to increase interest in STEM fields and expand the employee pool for Qualcomm. "It's taken on, in a very wonderful way, a life of its own," said Susie.

In the summer, Qualcomm offers Qcamp, a two-week camp for middle-school girls. One 6th grade girl wrote, "I learned that just because I'm a girl doesn't mean I can't. You know, I can do anything, and I'm smart, and I'm cool."

"Qualcomm is not in the business of education," said Susie. "Our interest is really in increasing the number of kids going into technology fields. And so, we'll be expanding nationally the same kinds of programs that have been so successful here."

Programs like Thinkabit Labs not only provide meaningful, real-world experiences for students but they also create connections between schools and businesses that can lead to additional collaboration.

Fictiv and Amador Valley High School: Pleasanton, California

The case of Amador Valley High School and Fictiv in the San Francisco East Bay is a powerful example of how collaboration between business and schools can provide benefits to both.

Students have real-world problems to solve and they gain practical experience; a local business has the ability to expand its network of contributors, benefiting not only their customers but also the local economy.

Fictiv is a hardware development firm dedicated to democratizing manufacturing by empowering people with the tools, information, and community necessary to build innovative products. They facilitate a more efficient production roadmap through a network of skilled professionals, resources, and tools to meet market demand. Fictive identified the market requirement: to make not only the tools necessary to respond to production needs available but also the talent needed to execute orders. In their search they identified an unlikely source of value: Amador Valley High School. Students are using the Fictiv platform to find work they can do in the school's Makerspace, as part of an engineering program. Students are designing and 3D printing parts for real customers.

Fictiv's CEO Dave Evans said: "This engineering program at Amador Valley High School is run by a very smart engineer. These kids are incredibly lucky as high school students to be learning about Solidworks (a CAD software program). And on top of it, he's teaching them through real examples and parts. Through this shared economy, this distributed manufacturing approach, all of a sudden, the win-win is that the high school gets to grow their engineering program, startups are getting their parts in 24 hours and innovating faster, and we are continuing to grow this ecosystem. That's game-changing."

• • •

Education for Everyone

A researcher in adult education at the University of Toronto, Allen Tough wrote a paper called **"The Iceberg of Informal Adult Learning**[8]." Tough formulated a reverse 20/80 rule for adult learning. Twenty percent of an adult learner's efforts were formal, organized by an institution. Eighty percent was informal, organized by the learner. He used the metaphor of an iceberg to describe the large portion of learning, informal learning, that remains invisible.

Tough researched the reasons why people chose to learn on their own rather than attend a class. "People seem to want to be in control," he wrote. "They want to set their own pace and use their own style of learning; they want to keep it flexible." Moreover, informal learning is often experiential and social. Lifelong learning organized around one's interests might be seen as a new form of recreation.

One of the drivers of the Maker movement is DIY learning driven by the internet and available to just about everyone. More people realize that they can customize their own learning, based on how they learn best. The internet is providing more access to learning tools, peer-to-peer interaction, and vast amounts of highly-detailed knowledge in all forms. It offers more variety in learning styles and outcomes.

That is why the future of education will not be found on a single campus. Instead the learner, more than the institution, will be in control, organizing their own learning path based on a combination of local and global resources. Yet, learners want to learn with others and people see value in learning with their peers.

A city should study how to increase the number of informal learning resources and promote them to new audiences. Informal learning requires motivated instructors, interested students, and available spaces. Organizations can play the role of matchmakers to increase the number of offerings. Potential teachers may need help recruiting enough students to make a workshop viable. Students may have difficulty locating a particular class at convenient time. A big challenge can be finding space in the community to host workshops and classes. The opportunity is to connect the various educational resources and make them more broadly accessible.

In short, much collaboration between the individuals and organizations that make up the learning community in the Maker City is needed. Find ways to get people with different perspectives together, using different ideas from the learning communities themselves. Then harvest those ideas into a set of experiments that make sense, ideally focusing in on a small geographical unit, working neighborhood by neighborhood to affect change.

STUDENTS AS AGENTS OF CHANGE

Students are agents of change in their communities, helping to create the kind of learning experiences that are most meaningful to them.

After completing the "Introduction to Drafting" class and visiting Maker Faire in Seattle, Ethan Toth, a student at Wenatchee High School in Washington state, decided to bring Maker Faire to his home town for his Senior Project. "I think what struck me at first was I love creating things," he explained. "I really saw the value in designing something with your own hands. However, I considered it more of a hobby... I couldn't believe it was a job... if I can use my skill set to convey this idea to everyone in our community, [that would be great]."

Wenatchee is a small town of about 30,000 on the Columbia River whose economic base has traditionally been agriculture and hydropower. The city was struggling to adapt to the technological and demographic changes happening around them, yet was positioned to benefit from open innovation. Ethan saw the potential that existed in his community; however, first the idea of the Maker movement had to be introduced to city officials.

"What I ended up doing was just going to local community leaders and asking if there was a way I could get involved in helping set up a Maker Faire... And it just generated enough excitement that our college, our city government, and the school districts all hopped behind it instantly and saw that creation really can inspire a community. The whole idea behind the Faire was to generate enough excitement for more permanent locations like a Makerspace or a co-working space to be placed within the community....We have Makerspaces starting to spring up in our local museums. The school districts have been getting stuff like CNC machines. And a lot of people, including a lot of my friends at the high school, are just inspired to continue on the Maker Faire."

The impact on Wenatchee has been remarkable.

Elected officials, including Mayor Frank Kuntz, became big supporters and proponents of the Maker movement and its role in revitalizing their community. Kuntz signed up as part of the White House **Mayor's Makers Challenge**[9] at a time when the word Maker was not lingua franca. Not only did it generate a more innovative, collaborative culture in Wenatchee, but it served as a unifier with East Wenatchee on the other side of the Columbia River. The two towns, despite being namesakes and being located so close, have had a history of civil and uncivil discourse.

"What is amazing is how this conversation [around Making] transcended political boundaries. It was something that everybody could identify with, and it was recognized for something that we needed," said Allison Williams with the City of Wenatchee. "It was beautiful to see everybody come together, understand the concept, and recognize it was something that we needed as a region to survive."

THE QUIET REVOLUTION HAPPENING TODAY INSIDE MAKER CITIES

The changes happening in education due to the Maker movement are happening at a grassroots level, driven by the passion and commitment of individuals and organizations at the forefront of formal learning and informal learning Inside their cities. At no time in history is there more potential to change: to turn passive learners into engaged residents of the Maker City, residents capable of thinking for themselves, using the tools of Making to rapidly iterate and prototype solutions. The Maker movement is unleashing a generation of passionate learners. Now we have people who are not afraid to pick up new tools and technologies and teach themselves how to use them, guided by a new generation of teachers who see themselves not as instructors but as coaches, coaches who guide their students to find their passions and apply what they have learned to real-world problems.

Cities have an enormous responsibility to prepare their young people for a lifetime of meaningful work. To the extent they succeed the students, in turn, take jobs in the public, private, or nonprofit sectors and/or work to create small businesses or freelance careers. To the extent the city fails in this endeavor, young people are left restless, disconnected, unemployed, and unemployable. No task that a city can take on is more daunting than education. Maker Cities recognize this and are engaged in a quiet revolution to transform education.

IMPLICATIONS FOR CITIES

Inspire true collaboration among different organizations in your city. An interest in Making can bring the entire learning community together and in so doing can reshape your city, turning it into a kind of open classroom that can serve the needs of not just young people but also lifelong learners.

Focus on the needs of K-12 learners. In the Maker City, learning should be student-centered and meet the needs of all learners regardless of age or income level. That said, it is important to focus on creating rich Making opportunities in K-12 schools settings. We believe that when children are introduced to Making at a young age, it pays dividends in unexpected ways by creating children with a lifelong love of learning, and the ability to problem solve and innovate with confidence, adaptability, and a sense of agency. More rigorous research is needed to quantify the impact of exposure to Making on not just educational attainment but other metrics of success.

Invest in champions. Your Maker City is likely to have a small number of champions with a strong interest in Making. By champions we mean the people in the learning community that care about creating more on-ramps to Making that are available to all learners, regardless of age or income level. Champions can come from anywhere in the learning community: teachers, principals, librarians, program leaders inside nonprofits, and youth directors inside churches and temples. Make it easy for champions to get together and share their interest in Making with others. Strive to create an open environment where champions for Making in the learning community can come together and share what works and what doesn't. Break down any silos that may exist that can get in the way of shared learning and collaboration.

Focus on organic growth versus a top-down approach. Administrators can nurture and grow the number of champions interested in Making inside a school system or other learning network, but they shouldn't take over the network in its entirety. A top-down approach is doomed to failure, largely because there is no "one-size-fits-all" approach to Making. Champions need to integrate Making at their sites in ways that work for their learners and their community, to allow the learning network to expand organically in a way that is sustainable.

Create an understanding and connection to Making that is deeply personal. Educators who are interested in integrating Making in the schools can start by reflecting on their own Maker core values. The goal should not be to fit Making into school-based learning; the goal should be to reframe and redefine schooling to allow Making, and students who make, to flourish both inside of school and out.

Give educators who are champions for Making time to learn, grow, and share. If you treat Making as a fad, then it will become just that. Change needs to happen on a number of different levels: pedagogical, curricular, organizational, and material. Get the required buy-in by conversing with administrators, parents, and other stakeholders. Making is not another thing to add to an educator's plate; rather, use Making to invite a cultural and curricular sea change. Time is required to develop a rationale for why Making matters at a given site.

Encourage educational champions in the Maker City to sign the Maker Promise. This is a simple and easy way to publicly surface your city's commitment to Making. Sign at makerpromise.org

Engage with corporations. Chevron has pledged $10M to the building of Fab Labs. Autodesk is providing free access to its software CAD tools for educators, and they have also supported Tinkercad, a free browser-based CAD tool that is popular with educators and students. Intel has distributed hundreds of its Arduino-based development boards to universities for free.

 Become an LRNG partner city. A great way to turn your city into a classroom, to encourage more "real world" learning. **A handbook for partners**[10] is available for download.

 Leverage students. The most underutilized resource in a school is its students. Rely on them as a barometer for feedback and insight.

RESOURCE

QUICK
WIN

IMPLEMENTATION
ADVICE

POLICY

BIG
IDEA

WORKFORCE & ECONOMIC DEVELOPMENT

HOW TO TURN THE MAKER CITY
INTO A MAGNET FOR TALENT

Maker Cities recognize that they must create jobs and opportunities to attract and retain talent in their cities. Without talent, the Maker City cannot hope to have the energy and vibrancy people expect when choosing where to make their living. As cities like San Francisco and New York, what some call gateway cities, become more expensive, a wide range of American urban centers are emerging as attractive and affordable options, both for individuals to live in and for companies to set up shop.

"More educated workers now leave Manhattan and Brooklyn for places like ... Cleveland and Buffalo ... than the other way around."

Source: **Daily Beast**[1], 2014

In short, there is a battle for talent inside our cities.

"One million young adults move each year, according to a 2014 City Observatory study. Their presence in a city is a direct reflection of its health and well-being, the study found, as young migrant professionals are key to fueling economic growth and urban revitalization."
Source: **Governing**[2], 2015

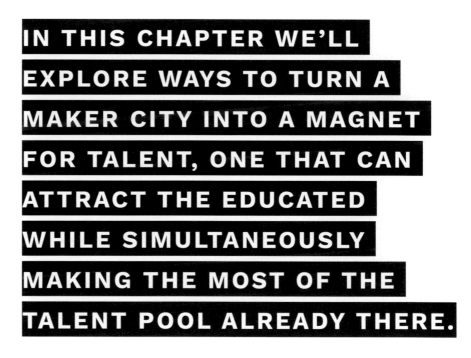

IN THIS CHAPTER WE'LL EXPLORE WAYS TO TURN A MAKER CITY INTO A MAGNET FOR TALENT, ONE THAT CAN ATTRACT THE EDUCATED WHILE SIMULTANEOUSLY MAKING THE MOST OF THE TALENT POOL ALREADY THERE.

One underlying theme here is that in the Maker City, formal credentials like a college degree are starting to give way to faster and more lightweight micro-credentialing programs that prove mastery of a specific skill or set of equipment.

Another underlying theme is the importance of training and apprenticeships to enable Makers to build portfolios that showcase skills and accomplishments in a concrete way. Building a strong portfolio is critical, enabling Makers to go after not just traditional work inside companies but also to create their own small businesses and/or hang out a shingle for freelance or gig work.

Exposure to Making → **Training/Apprenticeship**

↓

Three parallel and equally valuable paths to economic opportunity for Makers.

Work at Established Company

Create Small Business

Make a Living as a Freelancer

This ability—to move fluidly from traditional work to a small business to freelance—is important to keep people productive across a lifetime of work.

In the Maker City new types of businesses are expected to bubble up as a direct result of the Maker movement:

→ **Digital fabrication using 3D printing, laser cutters, and related technologies.** Expect to see these types of firms cropping up all over, much like copier shops, to revolutionize health care (prosthetics) and dentistry (braces, retainers).

→ **Custom fabricators.** Furniture and household goods built on spec.

→ **Makerspaces.** These will become increasingly professionalized and home to new businesses, and new learning/opportunities.

- → **Training programs.** Focusing on "understandings" and "makings," as explained later in this chapter.

- → **Job shops.** Leveraging 3D printers and CNC machine tools to do short-run, on-demand manufacturing. Also for product development: IoT (internet of things) and robotics.

Of course if these types of businesses are to achieve their potential, we need a cadre of trained Makers available in our cities with the right Maker skills and the capability of creating businesses from scratch, then scaling them.

Maker Skills Currently in Demand Include:

- → 3D Printing and Scanning

- → CAD/CAM and Graphics

- → Electronics and Robotics

- → Machine Shop and CNC

- → Plastics and Composites

- → Sewing and Textiles

- → Programming and Coding

- → Welding and Fabrication

- → Wood Shop and CNC

- → Fluid Power Systems

Andrew Coy is a former Executive Director of the Digital Harbor Foundation, a youth-oriented Makerspace in Baltimore, Maryland; today he serves as Senior Advisor at the White House coordinating Maker activity across the U.S. He notes that:

"We have too many jobs to fill to rely on one or two percent of the population that's just naturally in an economic position to tinker. Unless we find better on-ramps, unless we find better ways to train workers—in both formal and informal ways–we are never going to have enough jobs or enough people to fill the current demand."

We believe that cities can use seven main mechanisms to create an environment where Maker talent can thrive:

1. Embrace independent work and self-employment.

2. Build skills by focusing on new forms of vocational education (VocEd).

3. Enlist community colleges to train the next generation of Makers.

4. Focus on "jobs in the middle" when matching Makers who are seeking traditional employment.

5. Create new forms of apprenticeships and internships around Making.

6. Look to Makerspaces as drivers in workforce development.

7. Shift to new forms of credentials.

1. EMBRACE INDEPENDENT WORK AND SELF EMPLOYMENT

Traditional jobs are not the only way to make a living in the Maker City. About 34 percent of the workforce in the U.S. consists of people who work on a freelance basis, accounting for $715B in economic activity each year. (Source: **Freelance Union, Elance/Odesk Study, 2014**[3]). Freelancers work full-time, part-time, and also moonlight.

Etsy: From Freelance Work to Micro Businesses

Etsy is a company based in Brooklyn that makes a marketplace connecting independent Makers of handmade goods, largely made in the U.S., to buyers all over the world.

Etsy surveyed 94,000 Makers who participate in their marketplace to better understand who they are and what motivated them to focus on selling their handmade goods through the Etsy marketplace. (Source: **Etsy Study 2015**[4])

2005	**852**	**35M+**
Founded	Employees	Items for sale
1.6M	**25M**	**$2.39B**
Active sellers	Active buyers	Annual gross merchandise sales in 2015

The study tells us that one attraction for Makers is autonomy: the ability to work independently on their own terms and to build an income stream that does not depend on a traditional employer.

While the majority of Etsy Makers are women working from home and looking for a way to supplement their family's income; other Makers who work through the Etsy marketplace want to ramp up production and establish small businesses with employees.

Etsy takes care of the e-commerce end of things today. In the future, Etsy will take care of the manufacturing end as it allows Makers in its ecosystem to scale up. There are more than 3,000 sellers worldwide who have been approved to work with outside manufacturing partners on Etsy; 86 percent are working with partners in their home country.

The work Etsy is doing underlines the important role cities can play in giving Makers the tools they need to scale their businesses up. From an **Etsy blog post (2013)**[5]:

"Rockford is a city of 150,000 people, located two hours west of Chicago. Formerly a manufacturing hub, its keystone employers have left the city, ushering in a wave of high unemployment. One of the benefits of having a strong manufacturing history, however, is that many residents already have skills in the arts (such as watch making and furniture making) that were once the backbone of the local economy. Mayor Morrissey is an enthusiastic advocate of giving Rockford residents the tools that they need to turn these skills into greater economic opportunities.

"A group of Etsy Admin went to Rockford in November to further explore the needs of the city and opportunities to work together. [They] were greeted by a diverse group of over 70 stakeholders who gathered to share their thoughts on how Etsy could affect local constituents.

"[This] visit led to a plan for Etsy and Rockford to co-create a Craft Entrepreneurship Curriculum, with Etsy's platform and market-place as the learning lab. The aim of the project is to teach people that if they have a craft skill, entrepreneurship and economic opportunity are within their reach. Starting in September, the curriculum will be taught by local Rockford teachers to a diverse range of students."

The company doesn't expect every student to become an Etsy seller, but rather that students will apply the skills they learn to any entrepreneurial path they want to follow.

This pilot program has the potential to be not just what Mayor Morrissey calls a "pathway to prosperity" for Rockford, but a blue-print for similar programs across the country and around the world.

Makers need training in how to scale up their businesses with a curriculum that addresses fundamental issues of entrepreneurship.

When Makers Scale Up—What Do They Need to Know?

→ How to source parts for a product locally or online

→ How to assemble or kit a product using local manufacturing

→ How to determine options for fulfillment

→ How to obtain financing from local sources

→ How to find the right mentors and gain good business advice

→ How to develop a solid business plan

Private companies, universities, and Makerspaces can play an important role here, helping Makers get answers to these and other questions.

2. BUILD SKILLS BY FOCUSING ON NEW FORMS OF VOCATIONAL EDUCATION

For years, vocational education had a bad name; it was where little Johnny, who could read but didn't seem on path to college, got warehoused. Not any more!

Today's Maker Cities are borrowing from the tech industry, which has seen a wellspring of training programs springing up to prepare people for jobs as UI/UX designers, front end web developers, data scientists, and the like. Examples include General Assembly, Code Academy, and Galvanize, to name only a few.

We have to find faster, more agile paths for our young people to get trained and into the economy. This isn't traditional workforce development exactly; it's something else.

Vocademy: VocEd Revisited and Built around Making

Vocademy is a Makerspace that is quite explicit about its curriculum and focus: to train Makers to go into industry by exposing them to the right tools, experiences, and projects.

Started by Gene Sherman in 2013 and based in Riverside, California, Vocademy's curriculum is designed as an alternative to college for some and as a faster route into productive Making for others. According to Gene:

"In today's society everyone is funneled into college, but as we know that is not the path for everyone."

"Some go to work, some go to trade schools, but often you find that part of being young is not knowing what you want to do for the rest of your life.

"Say you think you want to be a welder. Great! You find a program that costs $12K and in six months you are a welder. What if three months in you hate welding and think CNC machining might be a better fit for your skill set? You've lost that time and you've lost some money. What we offer is like a gym membership, but this is a gym where the treadmills have been replaced with traditional and state-of-the art industrial arts equipment and tools."

Membership gains you access to tools and space and classes that help you develop basic skills and expertise. Vocademy is a Maker's dream. The 15,000-square foot Makerspace is filled with 3D printers, a full wood shop, laser cutters, vacuum former, plastics, traditional and CNC machining, welding, CAD, and sewing.

"The shop areas as well as the classrooms are all air-conditioned," says Gene. "I want to show people that manufacturing isn't all dirty and dark like you see in the action movies. Forty percent of our current members are women, a number I think will continue to grow as more and more people join."

Membership is only $99 per month and is open to anyone 14 years old and up. Minors need a parent on the premises, but daily operating hours are an amazing 10am to 10pm seven days a week to meet everyone's schedules. Typical classes range from six to twelve people, are taught by an instructor who specializes in that field, and cost only a few hundred bucks.

One of their more popular classes is the 40-hour "machine shop skill set." It costs $1500 and includes raw materials, introduction to cutters/abrasives, blueprint reading, math/measurement, and 20 hours hands-on with the Bridgeport manual lathe and mills. The advanced course moves on to CNC milling and turning via Vocademy's brand new Prototrak mill and lathe. "Southwestern Industries will be our CNC machine partners for life," exclaims Gene. "Their machines are perfect for what we are doing, the students love how easy they are to learn and work on. They are an amazing company to work with." (Source: **CNC West, 2016**)[6]

While Vocademy started out as a Makerspace focused on young people, particularly those not on a path to college, a curious thing happened. People of all ages and walks of life started to join as members and take courses. This includes older workers who had been edged out of aerospace and other industries in sharp decline in Orange County, the area around Riverside. Industry got involved to sponsor the classes as a way to fill critical openings.

Impact on Schools, Engineering Jobs, and Local Manufacturing

Today, Vocademy has a staff about 30 people and a membership of 250 people. Gene realized early on that community membership was not sufficient to sustain Vocademy. He began developing relationships with schools and businesses in the community. Two large classrooms at Vocademy are used during the day by the Alta Vista Charter School, which serves high school dropouts. The students come to regular class sessions but they can also take advantage of the Vocademy's hands-on classes, which they can take for free because of funding from the school district. "What attracts kids to Making is that it's fun," said Gene. "Our job is to make sure they acquire the skills of Making, which are invaluable." The charter school serves 100 students and 30 or more have taken classes at Vocademy.

Local businesses also use the space and its instructors to help employees learn new skills such as CAD software. In some cases, Vocademy helps engineers acquire the practical skills they weren't taught in college. One local manufacturer who had come to a Vocademy open house wrote Gene to say that his machinists were close to retirement and he would have trouble finding people to replace them. "I have good jobs at good pay," he told him. He wondered if Gene could recommend students for the jobs. Gene likes to talk about the skills gap and he believes Vocademy, as a place and as a concept, can help address it by taking Vocademy nationwide. He is actively looking for partners to expand his business model to hundreds of new locations and helping others open turnkey Makerspaces.

Riverside Mayor Rusty Bailey has been very supportive of Vocademy and declared in November 2013 that Riverside was a Maker City. Congressman Mark Takano said that a visit to Vocademy, which is in his district," sparked my interest in the Maker movement." Takano became a founding member of the Congressional Maker Caucus. Takano wrote on Medium: "As the Maker movement continues to spread, becoming an important part of American business and academia, it is drafting a blueprint for rebuilding our manufacturing base and creating a sustainable new sector in our economy."

3. ENLIST COMMUNITY COLLEGES

According to the Georgetown Center on Education and the **Workforce**[7], by 2020 65 percent of all U.S. jobs will require some form of degree or credential that goes beyond high school. About 45 percent of students enrolled in higher education in the U.S. do so through community colleges. Increasingly, businesses and policymakers are turning to community colleges to help fill workforce gaps.

In 2016 the California Council on Science and Technology issued a report and **call to action**[8] to growing a statewide network of Makerspaces linked to California Community Colleges as a key partner in developing a workforce for the innovation economy. The California system serves 2.1M students; where California goes, we expect other states to follow.

LCCC and Its FabLab

Lorain County Community College (LCCC) is based in Elyria, Ohio and serves 13,000 students per year. LCCC has honed in on Making as a way to ready its students for jobs in industry. A key initiative here is the college's FabLab, a kind of Makerspace that started at MIT and has spread, thanks to the work of the Fab Foundation, to 1,000 locations in 78 countries.

LCCC's FabLab is under the leadership of director Kelly Zelesnik; Kelly also serves as Academic Dean of Engineering, Business, and Information Technologies for the college.

LCCC FabLab

"One of the things that I appreciated about the FabLab as a former engineer and engineering manager is that I worked in an R&D environment for a medical device manufacturer. Quite often what would limit our ability to make progress on a project was we designed something, it was made, but it couldn't be completely made in-house. We would get some parts in and we'd put them together and find out there was a problem.

"And then we would have to turn around and resend it out, and we might have to wait for a week or two. It kind of slows down that design process. But when you have the ability to use all kinds of tools, both Making and digital tools like our digital FabLab, and you can turn something around quickly, it really allows you to design and iterate and get to that first article of inspection or the first prototype fast.

"As an engineer, I would have given my right arm to have access to a digital fablab. So to see that as an opportunity for students in our educational environment, not to mention our entrepreneurs, is very exciting for a few reasons."

Kelly then goes on to cite exactly why she is excited about the FabLab at LCCC. It turns out highly motivated students. Students return to the lab multiple times and log in additional hours, as needed, to really perfect their projects. The FabLab also gives students a way to build a portfolio of completed projects so as to gain skills and self-confidence and give them project-based work they can showcase to a potential employer.

4. FOCUS ON JOBS IN THE MIDDLE

Jobs in the middle aren't blue collar jobs, nor are they jobs for knowledge workers or high-level decision makers. They're something in the middle and a pathway into the middle class.

BLUE COLLAR | OVER 50% OF JOBS ARE HERE | DECISION MAKERS

Jobs in the middle used to include jobs in health care, jobs in construction, and clerical jobs. (Source: **Urban Institute, 2007**[9]) More recently, researchers at the Brookings Institute have focused on jobs in the middle in urban manufacturing, the skilled crafts and trades, as well as health care services and devices.

We know from Brookings and others that these jobs in the middle account for perhaps as much as 50 percent of job vacancies inside our cities. (Source: **Hozler & Lerman, Brookings, 2009**[10])

To succeed and thrive as a Maker City, cities need to create pathways for Makers to build the skills and display the competencies required for jobs in the middle.

Kelley Kline is Economic Development Director/Chief Innovation Officer in Fremont, California. Fremont is a town with an explicit manufacturing focus; it is home to Tesla among other factories and—as discussed in Chapter 6—thinks of itself as where Silicon Valley hardware products get built. Kelly articulated to us the acuity of the middle skills problem and how this problem is central to the city's economic development issues.

"We talk about the middle job syndrome. Some of this may be a little bit of a Silicon Valley problem in that we have a barbell economy, but we also have a barbell workforce. We don't have a lack of people at entry-level, and we don't have a lack of these academic people on steroids with PhDs. We don't have a lot of parents telling their children to aim for the middle. That's not happening. And that's where companies are really struggling.

"How do we get the dependable people in the middle that manage the line? Especially in really competitive environments where maybe some of those more capable, generalist people have a lot of attractive options, and the manufacturing companies have to compete for them.

"Veterans are a prime population for these middle jobs: they have qualities of leadership, and just basic know how in how to be a good employee kind of skills: be dependable and work hard, and have the right attitude. Tesla absolutely loves hiring veterans for all those reasons. Yes we have to train them what to do, but they will have the right attitude going into it, and that part is tough to train.

"Today the hardware part of the economy is starting to realize that they need to get kids excited about manufacturing earlier so that they can be more competitive. Kids are making a decision such as, 'Yeah, I don't want to sit in front of a computer all day. I actually want to be involved in making something tangible. And that's really cool, and that's how I want to spend my time.'

"In many ways, creating a Maker mindset is probably one of the more important things that we can do in terms of building the pipeline. The community colleges have struggled because employer needs are changing so fast. And to make curriculum changes at a community college even, which is the easiest level to make changes at, it can take years. They're always behind. They try really hard. And all of them now at least have industry councils that are advising them on making tweaks to have some of their programs address certain skill needs. But it just always seems to fall a little bit short. The thing I think has the most promise is to do what Europe has done so well, which is to really develop more of an apprenticeship system.

" A lot of the technology that you're seeing now within these advanced manufacturing environments is similar and people need to be exposed to it. It's just really hard to do that within a community college system. There's no substitute for actually doing some of this; actually being on a factory floor and having a chance to use some of this stuff. That's where people are going to get some of the key concepts they need for work.

Workforce Development is Changing in the Maker City

Many workforce development initiatives seem anachronistic, holdovers from an earlier (and simpler) time. Refreshing these programs is valuable, for at least two reasons:

→ **There's a real need and there's funding for addressing it.** In 2011 the federal government spent $8 billion on workforce development, much of it targeted to vulnerable populations (young people, handicapped, Native Americans) as well as to workers dislocated by changes in selective industries like automobile manufacturing that ended up gutting the middle class. (Source: **GAO Report**[11])

→ **Company retention.** Cities need a way to fill jobs that, if left vacant, might cause a major employer to leave a city.

Job-matching programs are designed to take the skills the unemployed have and match them to the job openings available through major employers. For job-matching to work you need to a have a well-developed pool of people with the requisite skills. But most of our cities no longer have this. In fact, there is every indication we are facing a skills gap, one that stands as a very real barrier to economic development in our cities.

Harvard Business Review looked at the issue in a 2014 article with the thought-provoking title, **"Employers aren't just whining. The skills gap is real."**[12]

According to Researcher Peter Cappelli, with the National Bureau of Economic Research, the problem is that "too many workers may be overeducated" or at least not armed with the specific skills needed today. The problem is in finding workers with the skills and inclination to fill the jobs in the middle, many of which take advantage of Maker skills. New forms of apprenticeship and internships can fill an essential gap here, providing skills in a manner contextual to work demands.

Shinola plant in Detroit

5. CREATE NEW FORMS OF APPRENTICESHIP AND INTERNSHIPS AROUND MAKING

Traditionally, workforce development was about on-the-job training. These kinds of programs have segued into on-the-job talent programs, recognizing that it's hard to make much headway by training people against a static set of skills. A more fluid, dynamic, and adaptive model is needed.

This is the approach taken by Shinola, a luxury goods manufacturer based in Detroit.

"It's one thing to invest in equipment and space," explained Jen Guarino, VP of Leather at Shinola. "I mean you can buy space, you can buy equipment. But talent is another thing. And so our huge investment has been more in training people how to do this work than it has been in equipment and space. That's easy stuff to just go buy or place an order on. You don't place an order on talent, right?"

Shinola needs workers with critical thinking that they can up-skill. To do that they've brought in the "masters," artisans and experts at their trades who have long retired, to come in and re-infuse American manufacturing with the expertise of quality production that was lost in a generation of workers who saw manufacturing exported.

This is a new form of apprenticeship, built around developing not one skill set or trade but around developing talent to work inside companies where making things by hand is an important part of the ethos.

YouthMade

Claire Michaels is Manufacturing Workforce and Hiring Manager for **SFMade**, which runs an internship program called YouthMade.

YouthMade gives low-income youth direct work experience inside small, urban manufacturing businesses. It is the first program of its kind, and it benefits both youth and employers: youth acquire transferrable skills and work experience while local business owners get to know this valuable talent pool better and nurture prospective employees.

The program started out small, with only 40 young people participating.

Adina Whitaker is a low-income student from the Mission district of San Francisco. Currently she attends San Francisco State University and works at Timbuk2, maker of custom messenger bags, as a YouthMade intern.

Adina started out in production work. She worked in customer service and even had a four-week opportunity to participate on the design team. The company then taught her inventory control, where she thrived and ultimately was responsible for redesigning the flow of the inventory management system in shipping and receiving.

YouthMade Intern working side-by-side with a more experienced worker.

According to Claire Michaels:

"This type of experience is invaluable in that it can expose people to careers they didn't know existed, to companies they may want to work with, and to the breadth of skills required to run a business."

79 percent of businesses involved in YouthMade said they would consider hiring their intern—assuming they had funds to do so—and 35 percent actually did.

The YouthMade program has since been replicated in New York through a partnership with Juma Ventures, SFMade, and the Southwest Brooklyn Industrial Development Corporation (SBIDC).

Towards New Forms of Apprenticeships
Apprenticeships have long been a means by which to transfer knowledge from an experienced craftsperson to a novice. Many private and public institutions are reinvesting in apprenticeships as a way of bridging the skills gap. In Detroit, JPMorgan Chase has made a $100M commitment to the city's economic recovery, investing in the Detroit Registered Apprenticeship Program and other workforce training initiatives.

Learn+Earn Apprenticeships
These are formalized programs, generally funded by the DOL (Department of Labor), that target underprivileged workers. Workers earn salaries during their training, which enables disadvantaged adults and young people to participate. The results of learn+earn apprenticeships are to expand economic opportunity.

As explained in a Brookings Institute study:

"Apprenticeship training culminates in career-related and portable credentials that are recognized and respected by employers. It relies mostly on learning in context, an effective method for teaching

technical and broader skills such as communication and problem solving. Although the U.S. apprenticeship system is small relative to systems in other countries, nearly 500,000 American workers are in the registered apprenticeship system and at least another 500,000 are in other apprenticeship programs."
(Source: Holzer and Lerman, Brookings Institute, 2009)

6. MAKERSPACES AS A DRIVER IN WORKFORCE DEVELOPMENT

Mark Hatch, Former CEO & Co-Founder of the chain of Makerspaces across the U.S. called TechShop, sees Makerspaces as an infrastructure play for cities.

Maker working at TechShop in San Francisco

Cities are tasked with the monumental responsibility of responding to unemployment and nurturing the workforce pipeline through high school graduation and college enrollment. Yet, 50 percent of high school graduates do not go to college. Making is increasingly an opportunity to serve as a bridging mechanism to engage youth and underemployed individuals around careers in manufacturing.

As Hatch sees it, "We still have trade schools. If you want to become a great world-class welder, you go to the trade school, do your 12 weeks or two years. But some of these tools are easy enough to use. You can actually bridge gaps very quickly."

TechShop was involved in one such program to "bridge the gap" that involved retraining union machinists in Pittsburgh, Pennsylvania. Union members were facing a layoff situation and needed to brush up on some of the modern machinery. They already knew parts of the work, but lacked the digital fluency needed to operate at a professional level with computer-numerically-controlled (CNC) equipment.

A new CNC water jet company was opening around the same time as the planned layoff; the machinists needed to be marketable to get the new work. After an 80-hour intensive training program, all twelve participants were able to secure jobs in less than a day.

TechShop now sees great opportunity in scaling this contemporary form of workforce development nationally.

In One Year, Doubling the Number of Patents Produced at Ford Motors

Ford has a number of fabrication facilities all over the globe. At the same time, fewer than five percent of employees are directly involved in designing cars or the parts that go into them.

Mark Hatch, Former CEO of TechShop and Bill Coughlin, CEO of Ford Global Technologies

Ford worked with TechShop to build a world-class Makerspace for Detroit. The idea was to expose workers at Ford to the skills and tools of Makers, so it could better compete in the innovation economy.

Bill Coughlin is a lawyer and the CEO of Ford Global Technologies, LLC where he handles IP and patents for Ford Motors.

"So I was trying to figure, how can we go from surfacing a great idea to prototyping just like IDEO and other such companies do. I was reading in the New York Times about TechShop and a light went off. This is it. This is the right thing to do because here's a group that has all the right machines for prototyping, has the classes, has the safety procedures, and they have this heart for Makers. So I gave them a call and said, 'You guys need to come to Detroit and I can help make that happen.' They couldn't believe it at first.

"The concept we came up with is giving Ford employees a free membership to TechShop for three months plus money for a couple of classes once they submit an invention disclosure to my team.

"So it became part of the invention incentive program within Ford Motor Company. What that did for TechShop was give them anchor tenants, if you will, of members. So I think we committed to 400, maybe it was 500 members, around the clock, annually. They could have that. I wasn't worried about it because I knew I had more than that in terms of Ford employees who want to invent. So they were able to come to start a new facility with relatively low risk."

In one year, the number of patents produced at Ford doubled. Plus, morale soared. People at Ford felt empowered to take their ideas on how to improve the cars made there and improve them through rapid prototyping.

7. SHIFT TO NEW FORMS OF CREDENTIALS

Portfolios are a way to establish that a Maker has not just the requisite skills but also the ability to work in a team setting when solving problems.

Making it in America Project

Bernie Lynch is the Project Manager for The New App for Making It In America, a $3M U.S. Department of Labor Workforce Innovation Fund project. According to Bernie, a big part of what is needed is competency-based skill training.

"It's not based on hours, it's based on your ability to perform certain competencies and demonstrate those, [to validate] that yes you have these skills and you're able to demonstrate that.

"Strangely, the DOL [Department of Labor] looked at the program, ran it through their skill dictionary, and found that it was like nothing they had ever seen.

"From our point of view, that was mission accomplished. We set out to make sure we were developing something that really meets the needs of Makers and the kind of training protocols that entailed. For the last 30 years in the U.S. we haven't developed the capacity for people to work with their hands to build products and businesses. That has to change.

"The way our model works is we break it into two categories we call understandings ['head'] and makings ['hand'].

"Understandings [are skills] around the notion of ... group

dynamics and working with groups, or do you know how to create a BOM, which is a bill of materials, which is a really important skill if you're going to learn how to scale your project and work with the rest of the infrastructure globally in making your products. So we teach a series of understandings of what is the entire process that you need to know to go from making one to making a million.

"Makings [are more technical skills] than the digital tool set. Have you learned the series of equipment, what are the materials, how many materials are you working with, as well as what are the methodologies and processes you need to know so that you're able to make things?

"Splitting the curriculum up in this way provides a flexible platform that enables people to come and describe the material they want to master, and the equipment to get them where they want to be, or, if they want employment, we can tell them, 'If you want employment to do X, these are the things you should learn.'"

Whether this type of non-credential skill development program will end up being offered through Makerspaces like Digital Harbor or Vocademy or community colleges like LCCC isn't clear at this point.

What is clear is that employers increasingly value skill-based training that reflects how work is really done inside industry today, where the focus is on cross-functional teams, problem solving, and rapid prototyping of potential solutions.

IMPLICATIONS FOR CITIES

Workforce development is a key part of the strategy for developing a Maker City ecosystem. Co-location of Makerspaces, educational institutions, and co-working spaces is a great way to build a community that shares skills which can provide a new kind of workforce development. Investor and financier Nathan Schwartz of Pittsburgh, Pennsylvania did this to good effect, using his family's real estate holdings to create a co-working space called Revv Oakland in close proximity to Carnegie Mellon University. This is an easy model for any college town to replicate. Find a local entrepreneur who has succeeded and encourage him or her to give back by setting up a co-working space for small business and Makers, ideally near a university or college. We saw a similar setup, albeit one that was sponsored more directly by the city itself, in Chicago's Hyde Park that now has an innovation center near the University of Chicago called Chicago Innovation Exchange.

Really look at local regulations and how they will affect people who freelance for a living, as well as emerging entrepreneurs. Make it easy for people to do business with your city. Consider streamlining the processes to apply for a business license, file a DBA ("doing business as"), or apply for a zoning variance.

Create training programs that enable people to gain practical experience in both "understandings" and "makings." An example of an understanding is crowd sourcing, taking advantage of platforms like Indiegogo, Kickstarter, and Fundable to help a Maker with startup ambitions get financing and figure out product-market fit in an accelerated time window. An example of makings is the type of training Vocademy is doing, to enable people to quickly and easily learn how to use machines like a 3D printer or a CNC-controlled lathe.

Create mentorships, apprenticeships, and internships. Mentorships/apprenticeships are being reinvented thanks to the Maker movement at companies like Shinola that encourage master Makers to transfer their knowledge of production techniques to Makers who are less experienced. Internships are likewise undergoing a sea change. High-school level interns can do highly productive work as we saw when we discussed Adina Whitaker, the intern working at Timbuk2 thanks to YouthMade. When structuring mentorships, apprenticeships, and internships move beyond skills training to application of skills in a work setting. The Department of Labor has funds available for Learn+Earn internships that enable disadvantaged populations to take on internships. DOL funds may also be available to target apprenticeships and internships at older workers who have been edged out of one career and now need to be retrained to enter another.

Think in sprints not marathons. A four-year degree program is a marathon that ends with a diploma. Think instead of sprints that end with completed projects; such sprints are a better fit with today's fast-moving economy. The examples of non-traditional workforce development programs we've cited here have built curriculums with courses that last anywhere from three to sixteen weeks, long enough for the Maker to find out if working with a CNC lathe is something they really want to do and short enough that the skills acquired are not immediately out of date. A system of micro credentialing could help here, to give Makers a way to gain validated recognition for the skills and competencies they've accrued.

Job matching may or may not work. As Bernie Lynch points out, one challenge is that the skills needed to succeed as a Maker defy easy categorization. The fast pace of business is another: by the time you train up a worker in a particular set of skills, the job may have disappeared. We saw this happen in the U.S. in 2008 when we tried to retrain construction workers for jobs in the solar industry. By the time training was complete, many of the jobs had moved offshore.

Makerspaces provide informal workforce development. Policy leaders like Andrew Coy believe that Makerspaces of all kinds are important to job creation. They can be large and freestanding, or smaller and embedded within libraries, schools, or rec centers. Wherever they exist, know that Makerspaces are enablers. They enable people to experience a 3D printer and CNC equipment for the first time, get skills-based training they need, and build a portfolio of completed work, one they can leverage into either full-time employment or freelance work. Making is deeply personal so it pays to invest in at least one or more Makerspaces that are large enough to include a range of equipment. If your city does not have the resources to do this on its own, consider teaming with TechShop or Makerbot, both of whom are adept at finding corporate sponsors.

ADVANCED MANUFACTURING & SUPPLY CHAIN

THE U.S. IS STARTING TO "THINK DIFFERENT" ABOUT ITS MANUFACTURING CAPABILITIES

Manufacturing is experiencing a kind of renaissance inside our cities, driven by changes in attitudes towards Making and changes in technology that enable small firms to produce high quality, high-value products and take advantage of emerging local and increasingly distributed supply chains.

The implications of this renaissance can and will be profound. Our competitiveness is on the rise as higher-value, more on-demand manufacturing becomes more common. Well-paying jobs are being created, often with new and more technical skills as a requirement. The converse is also true: older, lower-skilled jobs are going away due to the rise of new forms of automation.

Along the way, industries such as fashion, furniture manufacturing, textile production, and even electronics are being reclaimed and reimagined in the Maker City.

Local economies in particular stand to gain from new forms of goods that are not mass-produced but instead made locally, in relatively small batches, often with advanced materials and customized to better fit what people truly want and need.

The economics for manufacturing in the U.S. has shifted to the point where manufacturing, particularly high-value manufacturing,

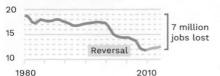

U.S. Employment in Manufacturing Industries (in Millions)

20 — 15 — 10 | 1980 ... 2010 | 7 million jobs lost | Reversal

Source: Brookings Institute, 2016

can be economically viable and happen "onshore" as opposed to "offshore." This is a marked departure from the 1990s when the United States effectively gave up most of its manufacturing prowess and process know-how to firms in China, the former Yugoslavia, and elsewhere.

In fact, urban manufacturing jobs in this country grew approximately ten percent each year between 2011–2015, after many years of decline. (Source: **Brookings Institute**[1], 2016.)

ASSEMBLY LINES AREN'T COMING BACK ANY TIME SOON

What is coming back is jobs in manufacturing that take advantage of advances in both hardware and software to add value. Smaller-scale manufacturing companies are also on the rise. The reasons for this are based on four distinct trends, according to the Deloitte Center for the Edge writing about the **Future of Manufacturing**[2] in 2015:

Source: Deloitte Consulting, 2015

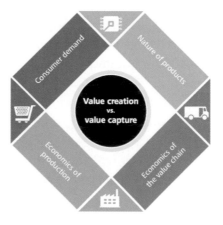

→ **Consumer tastes are changing.** Luxury goods. Niche products. Limited Editions. Products created as platforms, that can be customized by the consumer or business buyer. All of these reflect the changing dynamics in tastes, changes that favor domestic production.

→ **Nature of products.** Products have stronger software, technology, and R&D inputs, which makes for more frequent development and update cycles. Hardware products are taking on characteristics we associate with web services: continuous change rather than defined long-term product cycles. This stresses overseas manufacturing and favors local supply chains. Rapid technology changes also mean we replace products more frequently. The plain old telephone (POTS) is the poster child for this; Americans used to purchase (or rent) a phone for ten years or longer. Now, we update our mobile phones much more often, with some people paying extra for the right to update their phones with each new model which can mean updating one to two times per year.

→ **Economics of production.** Both the software and hardware used to support the manufacturing process have evolved to the point that they have dramatically changed the economics of production.

→ **Economics of the value chain.** It used to be that much of the value in manufacturing came from the assembly process itself. Today, much of the value comes from anything but assembly, from advanced materials to novel production processes. Increasingly, manufacturers are delegating the nitty gritty of assembly to a network of suppliers, controlled by cloud-based software and manufactured using a network of loosely coupled equipment and operators available on demand. (Think of this like Uber, but for manufacturing. Why own a factory if you can manufacture your product on demand using other people's software and hardware?)

The Importance of Clustering around our Strengths

These factors are leading to a resurgence in manufacturing of all types, shapes, and sizes in U.S. cities. It is still the case that manufacturing capabilities inside U.S. cities are spread out, without the density one sees in a city like Shenzhen, China.

To build lasting competitive advantages, we believe it is important to pay attention to density, to build manufacturing resources not just in a city but also in the surrounding region. Regional clustering is important to enable fast cycle time so that Makers can create a design, see it produced, and iterate to perfect that design for its intended use and for manufacturability.

This idea, that clustering leads to competitive advantage, is not a new one. It's been around since at least 2009 based on the pioneering work by Professor Michael Porter at the **Harvard Business School**[3].

In their book, The Smartest Places on Earth: Why Rustbelts are the Emerging Hotspots of Global Innovation, authors van Agtmael and Bakker (2016) argue that the era of cheap is over and that we now compete in the era of smart.

Likewise, Deloitte believes that the kind of **advanced manufacturing**[4] the U.S. needs to develop in order to build lasting competitiveness will come from relying on predictive analytics, the internet of things (sensors), and advanced materials. In other words, the products we manufacture here will have relatively complex technology inputs.

Keep this in mind as we talk through the examples below. At the end of the chapter, we'll talk about the future of advanced manufacturing inside the Maker City.

CHANGES IN CONSUMER TASTES

Today, when consumers show a preference for products that are Made in America, they do so for quality reasons, but also because products made locally can meet the specific tastes and needs of U.S. customers in ways never before possible.

As a result, Makers are stepping in to invent niche products, posting them on crowdsourcing sites to judge demand in advance, and then building the products in micro factories established inside the Maker City.

The Rise of Micro Factories: Nomiku Wifi Sous Vide Cooking Device

A micro factory is just what it sounds like: a complete factory executed in a relatively small footprint and run by only a handful of people thanks to the fact that most of the machines are computer controlled.

Lisa and Abe Fetterman were determined to try making the Wifi Nomiku, the world's first wifi-connected sous vide device in the United States, so they moved back to San Francisco.

"In China, we set up our own line, and engineered our line. We had to build our own things anyway," said Lisa. "The first 100 Nomikus were touched by Abe." They lived right next to the factory. "It was the only thing we did," said Lisa. "We lived in basically dead farmland."

She found it exasperating. "You have to be there for every single step and watch. You trust but then you have to verify everything, step by step, all the way. We thought if we have to do this anyway and it takes so much time, why can't we do it in the United States?"

While Lisa continued working on marketing and sales, Abe began researching how to set up their own production line. Trained as an astrophysicist, Abe learned to solder at a Makerspace, practiced by building kits, and eventually felt confident enough in his soldering skills to set up an assembly line with workers who would do the soldering for him.

While he was in China at the factory, Abe saw opportunities to do things better, more efficiently. "In the back of my head, I was always questioning 'why am I here if I can't make things better?'" He had read articles about how the production system in China is so great because you can go visit the factories. "They said everything was within like two hours," he told us.

"But then, you realize, that's a nice theory. The truth is it's not like anything is less than two hours away. There's so much traffic, it's hard to get around.

"Once we needed a specific washer, and were told that nobody makes that washer within two hours of the factory. So we took a

Sous vide is a method of cooking vacuum-sealed food in a controlled low-temperature water bath. With sous vide cooking, there is no need to remain tied to the stove; the cook gets notified when the food is ready. The process itself retains both the flavor and nutritional value of the food.

boat to Hong Kong, found a shop that had the washer, bought the washer, took the boat back, and then brought it to the factory, because there was no more efficient way to get that washer. That's a ridiculous thing to have to do in order to get your product made. If I was in the U.S., I could go on the internet and order it, and have the part delivered overnight.

"One thing I took away from China is that people there are really good at making things that they have already made, but you have to be the expert in making your product [if it's new and different]. Nobody else is going to be the expert in that. So what our goal has been is to rely on China to make things like our pump heaters. They make millions of them. But nobody makes Nomiku. By doing it on our own, we have more opportunity to take advantage of new processes and new ideas."

Abe also wanted the engineers and sales teams to be in the same space so they could work together closely. He had seen how MakerBot had its engineering in China and everybody else back in Brooklyn and the two teams had trouble syncing up. Instead, each group blamed the other one for delays. "You need to have your engineers in the factory or else you have this disconnect where you don't know what's going on and you're not controlling it and someone else is making your product. You shouldn't do that if you're a startup. Your entire company depends on that product."

The most expensive part in the Nomiku is the printed circuit board—its electronic brain. They ended up having that made in China. The plastic casing and other parts required tooling for injection molding. They ordered a tool that would be made in Taiwan and then shipped to their San Francisco office.

At the end of 2015, Nomiku started with five people on their assembly line and that number might grow to ten. Abe said that it was hard finding people in San Francisco to work in production. "People aren't really used to doing this in San Francisco. It just seems that turnover is high or you have to pay a lot."

It wasn't easy but Lisa and Abe now have a small factory ready to make a sous vide cooker in the U.S.

"When we started to manufacture in the U.S. we said, why not?" said Lisa. "We're already so crazy for doing all this stuff, it just adds another layer of crazy. Let's see what happens." (Source: Dale Dougherty, Free to Make, Fall 2016)

Lisa and Abe used the crowdfunding site Kickstarter in 2014 to raise over $750K from 5,538 backers; another entrant in the sous vide category launched just before they did on Kickstarter and raised over $1.8M from 10,508 backers.

How Crowdfunding Works

Crowdfunding is a win for both Makers and consumers. Makers get immediate feedback on whether their product ideas have merit. Consumers get early access to unique products that are typically not available in stores.

→ Maker gets idea for product.

→ Maker creates a prototype and figures out whether the prototype can be manufactured, perhaps using a system like Plethora (discussed below).

→ Maker creates a video explaining the product, its benefits, and why the product can be manufactured

→ Video gets posted on a crowdsourcing site. Indiegogo and Kickstarter are two examples. With crowdfunding, the community decides what products get made, by voting to back a particular product. Many of the products promoted on these sites are little more than prototypes.

→ Product reaches its funding goal and gets funded. With funding comes the obligation to move from prototype into manufacturing mode.

→ Product gets manufactured, possibly using a system like Fictiv (discussed below) to source component parts, and ships to consumers in a matter of months. Note that some more complex products can take years and/or may never make it to consumers.

CHANGES IN THE NATURE OF PRODUCTS

Products are becoming platforms, co-produced with the consumer, taking advantage of crowdsourcing in a different way, one that encourages Makers to create products in response to a crowd-sourced challenge.

GE FirstBuild, based in Louisville, Kentucky, operates as a joint venture between General Electric (#8 on the Fortune 500 list) and Local Motors, a company you've probably never heard of.

According to Venkat Venkatakrishnan, Director of FirstBuild, GE typically spent $40M-$50M to create a new consumer appliance which took about four years. So lengthy was this cycle time that by the time the new appliance came to market it had only a 50 percent chance of success.

To cut spending and reduce cycle time, FirstBuild relies on an open innovation model which encourages Makers and others to submit their designs in response to a specific challenge.

For example, one challenge asked Makers to submit their designs for a cold-brewed coffee machine. Designs are submitted to FirstBuild, which reviews them; acceptable designs are posted to a crowd-sourcing site called Indiegogo where consumers can pledge to fund the project and get access to the "first build" of the product.

"Design the aesthetic and interaction of a device which cold brews coffee in minutes rather than hours."

Dr. Anthony Townsend is a Senior Research Scientist at New York University's Rudin Center for Transportation Policy and Management. He talks about a virtuous circle as a kind of "well-functioning cultural production system."

"One that starts by figuring out trends, what people like, and what is authentic, then it packages all of this up in neat ways, then figures out how to make it cost-effectively, then determines how to market it to people, and get it to them. And it's working."

This virtuous circle enables FirstBuild to co-create new consumer appliances with their consumers. It has cut the time required to build a new consumer product from an average of four years to an average of four weeks. Creating a typical new product now costs only $50K as opposed to $40M -$50M. Most of this money is spent on a promotional video posted to Indiegogo and for prizes that go to the winning design team(s). Design teams can be made up of consumers with very little specialized knowledge, just the willingness to "hack" an outdated appliance to make it better fit their needs. Because consumers vote with their wallets, FirstBuild knows that the winning products meet consumers' needs.

The products that come out of FirstBuild are low-volume projects compared to mainline GE appliances; they are experiments in production that yield invaluable market knowledge, considerably reduce new product risk, and school the appliance division in manufacturing innovation.

So successful is the FirstBuild system that it has extended this co-creation model to the design of the parts that get built into its appliances. When FirstBuild needs to create a new part for one of its consumer appliances, it puts out a challenge to engineering students at the University of Louisville who often produce responses in a matter of hours with designs for the component parts. The winning design gets funded and often FirstBuild brings the student into its lab to refine the design, so as to smooth out any hiccups in the manufacturing process.

The meta problem that FirstBuild solves for GE is that a brand that has traditionally appealed to late adopters, is now being helped by and appealing to early adopters. Thanks to FirstBuild, GE now has access to a steady flow of ideas and innovations that can lead to entirely new product categories plus the ability to prototype these ideas and manufacture them onsite, meeting the immediate demands of early adopters before they move on to the next great thing.

When you visit FirstBuild, you are struck by the sense that something new and radical is happening there. Entering the premises is almost like walking onto a giant soundstage with different sets, each for a unique scene in a motion picture. To the left is a "showroom" of hacked, one off, and prototype appliances. Inside the showroom is an oven, which has been (safely) modified to become a 900-degree pizza oven by having its self-cleaning safety mechanism safely modified.

"Our customers kept doing this, so when a team of cooks and Making enthusiasts came to First Build to do it themselves, we watched them, helped them, and then helped bring this thing we'd never thought of to market," explained Mr. Venkatakrishnan.

Next to the pizza oven is an oven in which the rack automatically slides out as the door is opened. Next to that stands a refrigerator with a raft of USB connectors inside so developers can write to the refrigerator as a platform. This is because the team at FirstBuild believes that an appliance should be viewed as a platform that consumers and developers can add to, much as is done with smart phones or computers. To the left of the appliances is a GE test kitchen that also serves as the front of a gathering place and auditorium for presentations and conversation.

Down the hall behind giant doors is the real action: a three-story tall 3D micro factory, one of the most complete Makerspaces we've seen.

GE is leveraging Maker technology to improve its flexibility in manufacturing, its responsiveness to consumer trends, and its ability to create unique IP for defensible advantage.

CHANGES IN THE ECONOMICS OF PRODUCTION

Manufacturing at Point of Sale

Thanks to 3D Printers, Makers are now inventing products to be manufactured at point of sale, which reduces inventory carrying costs and end-of-season returns.

Aly Khalifa is the founder of LYF Shoes based in Raleigh, North Carolina. Lyf shoes is an experiment in how technology can be applied to create completely customized shoes, built for the exact dimensions of each of your feet, taking into account that your left foot and your right foot have different measurements.

"If you want to talk seriously about re-shoring production in the United States, then you have to talk about completely changing the game. This means designing for decentralized manufacturing," says Mr. Khalifa.

Khalifa envisions distributing manufacturing as a way to significantly reduce inventory carrying costs for shoe retailers. "For a $250 shoe, a retailer would need to stock $6K in inventory for one style in three colors. Also, an enormous amount of cost in fashion goes to markdown. In the shoe industry, you might lose 40 percent of the shoe's value to markdowns just because you didn't sell a particular size in a particular color that season."

LYF Shoes production process

LYF sees the cost of industrial 3D printers coming down to the point where it is practical to manufacture customized shoes at point-of-sale so as to reduce inventory-carrying costs. Right now for LYF, "point of sale" means over the internet but could eventually mean turning local shoe retailers into manufacturers.

LYF takes advantage of American's desires to purchase a product that is fully customized to their needs versus mass-produced products, which are "one size fits all." In this, LYF is a lot like Timbuk2, a maker of messenger and other bags based in San Francisco, California. Consumers select their options through an ecommerce site, submit their order, and in a matter of days the product arrives.

Ironically, urban manufacturing firms can charge more for a highly customized product. Kate Sofis, of SFMade, explains:

"From a supply chain point of view they can; the minute you start customizing anything you can significantly mark up the price beyond the actual real incremental cost. It costs Timbuk2 no more to make a bag with customized panel colors than it does to make a bag where you haven't specified colors and they're just doing it to stock. Yet they can charge more."

Higher prices for customized goods can be offset by taking the cost of assembling the product and shifting that burden to the consumer. Consumers can either save some money by assembling the product themselves in the privacy of their own living room or work with Task Rabbit or a similar service to pay someone to assemble the product for them. Either way, with this type of distributed, highly-customized manufacturing there's no assembly line. As the U.S. looks to onshore manufacturing, its focus is on higher-value products rather than on the kind of low-value, mass-produced products that are better manufactured offshore.

Integrated Making allows designers the test and learn more efficiently; iterating prototypes rapidly and in the field.

Precision Manufacturing Using 3D Printers

Another way the economics of production have shifted is due to low-cost, fast, and highly precise 3D printers. This enables industrial concerns to iterate more quickly, as needed, to optimize a part or component for a particular use case.

Industrial-class 3D printers enable industry to move beyond merely prototyping a part but to actually manufacturing that part locally. This cuts shipping time and costs, and supports rapid iteration and optimization. This is having a transformative effect on the manufacturing prowess of this country.

Aerospace in Portland, Oregon

Patrick Dunne, Director of Industrial Applications for 3D Systems and a maker of 3D printers, gave us one example.

An (unnamed) aerospace manufacturer based in Portland, Oregon is using 3D printing to rapidly iterate and optimize a critical part, iteration and optimization that can only happen when a 3D printer is fast enough to support numerous iterations in a very short period of time. This aerospace manufacturer accrues a significant competitive advantage through a better, more energy-efficient part, one that is two percent more energy efficient than the old. While two percent doesn't sound like much, it matters a lot to the airline carrier that will buy and operate the plane on highly competitive routes. Moreover, the aerospace manufacturer is building up process know how. This creates a virtuous circle:

Other industries that can benefit from this virtuous circle include health care (prosthetics), transportation, energy production (power plant optimization), and aerospace. In short, any industry that depends on highly precise component parts.

Custom Prosthetics via an Open Source model

Another application where it pays to be highly precise and to iterate multiple times is in the creation of custom prosthetics.

A sculptor and designer in Paris, Gael Langevin bought a 3D printer for his work studio. He saw the potential of 3D printing to create objects for his commercial customers. "Some people think 3D printing could only produce little rabbits or Yoda figures," said Langevin "but I had in mind that it could be used in a more practical and engineering way."

A French car company asked him to bid on a futuristic prosthetic. However, even when the job didn't happen, Langevin decided to design the prosthetic hand anyway in his free time. "I always liked hands, my workshop is full with hands, made in all kinds of materials, some were molded with plaster on my own hands when I was 12 years old," he said.

After designing and printing the first hand, he decided to post the digital file on Thingiverse in January 2012 under a CC-BY-NC license, a type of license popular with the Open Source community.

"Sharing the parts with the Open Source community was a logical route for me."

The first hand he posted on Thingiverse in January 2012 was the first "possible" prosthetic hand design—a 3D printable hand actuated with tendons and motors using Arduino.

Since then, many 3D prosthetic hand designs have been released over the internet, based on Langevin's design or just inspired by his work. BionicoHand, RoboHand, OpenBionics are some of them. Langevin continued on to design an open-source 3D printed robot called InMoov.

There are communities sharing 3D printed designs and helping not only to improve the designs and customize them but also helping to make the prosthetics for others. Jon Schull from Rochester, NY started the **E-Nable Foundation**[1] to help coordinate the activity of a worldwide community of volunteers. The amazing thing is how the technology and a community willing to share have empowered people to create custom prosthetics that they could not afford to buy, or didn't even exist at all.

We see this work at Maker Faire. A father from Cincinnati taught himself to use a 3D printer to create a custom prosthetic hand for his young son. His son had chosen colors and a Star Wars logo for the hand. A group from Calgary called Make Fashion incorporated 3D printed prosthetics into a fashion show where women were able to wear prosthetic legs that were not just functional, but beautiful. Lisa Marie Wiley, a veteran who lost her lower left leg in an IED explosion in Afghanistan, was uncomfortable with the fit and feel of the prosthetic limb provided to her. As a petite woman, she realized that the prosthetic was not designed for women, especially not for someone her size. She was able to work with others to design and print a prosthetic that really worked for her.

FROM	TO	EXAMPLE
Hardware	Software	Plethora Fictiv BriteHub
Assets	Services	Plethora Fictiv Maker's Row BriteHub
Longer Production Runs	Shorter Production Runs	Flex Maker's Row
Focus on Reducing Cost	Focus on Reducing Risk Focus on Increasing Flexibility	Flex

CHANGES IN THE ECONOMICS OF THE VALUE CHAIN

It used to be that manufacturing was limited to those who could afford to rent a large space, then purchase and set up a slew of complicated, big, and expensive equipment. This is no longer the case in the Maker City. In fact, much of the value chain in manufacturing is shifting.

Manufacturing-as-a-Service

Imagine a virtualized factory that is less defined by specific hardware; but rather defined by cloud-based software that runs hardware owned and operated by others, connected together into a loosely coupled network. This model is called "Manufacturing-as-a-Service."

Traditionally, in a software-as-a-service (SaaS) company like Salesforce you "rented" the software you needed on a monthly basis versus owning a piece of software that you installed on your own hardware.

Companies that compete in the Manufacturing-as-a-Service segment take the SaaS business model a step further. Not only do you rent the software capacity when you need it but also you tap into the logistics and equipment network these companies provide, potentially eliminating the need to set up your own factory or lines of production.

This makes "Manufacturing-as-a-Service" more like Uber or Lyft, the popular car-sharing services. Instead of purchasing, maintaining, and operating a personal automobile, someone who needs transportation can depend on a logistical network in the cloud that routes a car and driver to their location on demand.

Similarly, with Manufacturing-as-a-Service, someone who wants to manufacture a product can depend on a logistical network that exists in the cloud to route their manufacturing request to the equipment and operators most suited to take on the work. Suitability can be determined based on a myriad of factors including: equipment, capacity, pricing, quality, and proximity to the buyer so as to reduce shipping costs/time.

This lowers the barrier to entry so that any Maker with a good idea and design prowess can become a manufacturer.

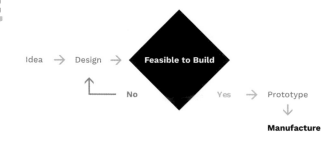

Plethora: The Network-Based Factory in the Cloud

Of course, first the part itself has to be designed in such a way that it can be manufactured with ease. Enter Plethora, a startup based in California that focuses on the design process.

Nick Pinkston is the CEO of Plethora, a company that wants to make creating a custom part from a design as easy as the push of a button.

Plethora operates as a plug-in to CAD/CAM tools like Autodesk. While learning CAD/CAM software can seem daunting, many Makers have at least a basic knowledge (if not mastery) of this type of software. A Maker can design a component part inside the Plethora system and get immediate feedback on the feasibility of that design from a manufacturing perspective. This cuts weeks or months out of the process required to take an idea from design to prototype.

If you are a product designer in San Francisco, you could sit at your computer and design a part, get feedback from Plethora's software as to whether the part can be made or not, and then press a button to send a design file over the internet to Pinkston's

Plethora factory, where it will be directed to a machine. After the part is machined, it will be sent by courier to you.

As a designer you'll have your part in your hands, possibly within hours. Pinkston believes that he can automate the order taking, the validation and testing, the interface that controls the machines, and the set-up for the job. If he's successful, his factory would be replicated in many places around the country or the world. If that happens, he would put many of the machinists currently working out of business, as well as the job shops they work in.

"There are 35,000 job shops in the United States. Who knows how many overseas? You go to those places. You call them on the phone. You send them your files. That's how it happens. It's basically just way slower and more expensive to do that. What we're trying to do is to bring the like Amazon Prime experience where you're just like 'Hit a button' and it shows up."

Why have your own machines or a machine shop when you can get access to one? Certainly, this has been part of Mark Hatch's idea about TechShop. You get access to any machine when you need it, for a monthly membership fee. Nick Pinkston is going one step further. You don't need to go to TechShop or a fablab to access local machines.

When the Plethora factory is scaled up, it will be full of sophisticated machinery but employ almost no one.

As Nick sees it, he's taking manufacturing to its logical conclusion by automating it. There aren't robots exactly in this factory; what happens is mostly in software that controls the machines. To be clear, it's not the ability to send design files to a manufacturer that is the innovation here. That has been happening for quite a while. One can send a design file directly to a factory in China, if it is prepared to do the work. Nor is it automation per se. It's relatively easy to automate a process to create the same object over and over again. It's much harder to automate a process when the objects are different with each job. It's flexible manufacturing. Pinkston's software makes flexible manufacturing a reality for more Makers, by dramatically bringing down the cost of retooling the line. (Source: Dale Dougherty, Free to Make, Fall 2016).

Fictiv: Compressing Cycle Time to the Bare Minimum

Dave Evans was one of the first hires at the Ford Innovation Center which the company decided to strategically locate, not in Detroit, Michigan but in Palo Alto, California. Dave spent a lot of time at Ford looking at the hardware development cycle and how best to compress it.

The development lifecycle for a car is four to five years, seven to eight years if you are talking about the more exotic ones. But if you think about consumer electronics, it's a life cycle measured in six to eight months, not years.

"So imagine, when you buy that 2016 Ford Fusion, the reason why that touchscreen feels like a second generation iPhone is because it is. It was developed in 2010, 2011, " said Dave.

So in 2013, Dave quit his job at Ford to found Fictiv with the idea of changing hardware manufacturing for the better, making it faster and more democratic in the process, so that anyone could design a part, see it produced, and iterate quickly, without first having to buy an industrial-strength 3D printer.

"There are other players that do this of course, like Shapeways, but none really are suitable for precision applications, where getting a part "right" requires that you iterate multiple times."

Fictiv excels at both fast turns and highly-precise manufacturing applications using 3D printers and other CNC machine tools (lathes, mills, cutters, and the like).

Before Fictiv, a designer would create a design and ship it off to get 3D printed or produced at a job shop with computer-controlled tools. It would typically take two to three weeks to get a part back. This doesn't sound like a lot of time—until you consider that it typically takes three iterations to get the part right.

Suddenly a process that was built on the premise of two to three weeks cycle time turns into a three-month process.

"Now, imagine, which is what we do, you get the part back in 24 hours. Then all of a sudden, you can do four revs in a single week.

What does that do to innovation, right? It completely changes the way that you build physical products."

Fictiv's secret sauce is software and a network that enables distributed manufacturing, where parts are intelligently routed to available machines. To the company's way of thinking, this results in faster parts, fair prices, and focused innovation.

Fictiv tags vendors based on their expertise, vets the contract manufacturing/job shops involved for both speed and quality, and catalogs them to make for a kind of virtual supply chain. Suddenly the supply chain that used to be available only to very big companies is available in a much more democratic fashion, to anyone with an idea and access to design skills.

SCALE UP THROUGH URBAN MANUFACTURING CENTERS

As a result of the changes in both the distribution and manufacturing processes, there is a resurgence of interest in manufacturing inside our cities. This is only possible because U.S. factories of today are smaller, employ fewer people, and are environmentally sensitive, so that they can be located adjacent to housing and schools.

Fremont: Tesla Motors

Fremont, California is famous as the home of the Tesla Factory, where the next generation of fully electric cars are manufactured using a combination of humans and robots.

Tesla built its factory on the site of a defunct GM assembly plant that closed in 1982 after 20 years of operations. The plant was reborn again in 1984 as NUMMI—a joint venture between GM and Toyota—only to be shut down permanently in 2009. Tesla purchased the land and facilities for pennies on the dollar in 2010.

Today, the Tesla Factory is the very picture of a modern factory at scale. The Tesla Factory started out employing 899 people and now employs 6,000+ people in 5.3M square feet, relying on robots to do the "heavy lifting" as well as the low-value assembly work.

Source: Tesla website. The Tesla Factory takes advantage of the $50M Schuler Press, which is the only hydraulic press in North America. It's seven stories tall and is responsible for stamping out Tesla's aluminum body panels.

Source: City of Fremont

Fremont has remade its city around the needs of factories like Tesla. In late 2010, news broke that 160 acres of vacant land surrounding the Tesla Factory was on the cusp of being acquired by Union Pacific Railroad. UPRR was determined to utilize the plot of land as a repair center for its trains and a distribution center for shipped goods. Neither use fit the city's vision of itself as a tech hub with a focus on advanced manufacturing.

Mayor Bill Harrison and other city leaders went to work, wasting no time in trying to reverse these less-than-ideal plans. What emerged from this crisis was an Innovation Zone purposefully built to allow advanced manufacturing firms to thrive in Fremont. The design and zoning of the city is mixed use, with housing and manufacturing plants located in close proximity to each other. Multi-modal transportation allows people to get around with ease. (Source: **Route Fifty**[5], 2016)

A similar transition is happening today in San Leandro, California which is a city of 85,000 about 24 miles north of Fremont. San Leandro has something that's at a premium in the Bay Area: space. When many of San Leandro's factories shut down in the '70s and '80s, they left behind underused space, places like The Gate, a former Plymouth auto plant that combines a Makerspace, a co-working space, and space for urban manufacturers who can start small–say in 1,000 square feet–and scale up to 10,000 square feet.

Once a manufacturing concern outgrows the space available at The Gate it can move into freestanding space. In 2015, Kraft announced it was closing its sprawling complex in San Leandro. City leadership is confident they can find a new tenant for the plant:

"We have been attracting more technology companies," said Jeff Kay, San Leandro's business development manager. "A lot of them have a focus on advanced manufacturing, the industrial side of tech manufacturing. (Source: **San Jose Mercury News, 2015**[6])

DEFINING THE FUTURE OF MANUFACTURING

If you read the popular and technology press, the future of manufacturing is coming in the form of robots that will replace humans on the assembly line. The future is almost certainly more nuanced than this, in no small part because robots are everywhere and have been for years. Robots include mechanical devices guided by AI and are fed data not only by sensors but also by software bots and process automation.

Robots provide an economical way for manufacturing firms to grow and reach their full potential in markets where human capital is prohibitively expensive and/or where there simply are not enough workers to meet demand.

The Tesla Factory illustrates this perfectly. Over 325,000 people put down deposits for the Tesla 3, an electric car priced at $35K (base) or $60K (fully loaded). To put this number in perspective, know that Tesla just finished selling approximately 50,000 of its electric vehicles in 2015, produced around 15,000 units in Q1 2016, and expects to ramp to 500,000 units by the year 2020. At Nissan's Sunderland factory in the UK, it took 28 years for the site to hit a production level of 500,000 units annually. (Source: **Inside EV, 2016**[7]).

This means that to meet existing demand, Tesla will need to hire humans and deploy lots of robots. Robotics itself is a growth industry for the U.S. Robots will enable manufacturing plants here to reach economies of scale earlier and with greater speed.

No one knows what the future holds and how—exactly—the rise of robotics will reshape the workforce and change factories. What we can tell you is that the future belongs to those cities that are relentless about experimenting with what works, iterate constantly, and display restless curiosity about how to deploy technology for the better.

SAN FRANCISCO: FLEX INNOVATION LAB

San Francisco is famous as home to thousands of startups, many of them founded by Makers intent on producing the next consumer appliance, consumer wearable, or drone. This has created a demand for manufacturing innovation, to bring design closer to manufacturing, and find ways to make small production runs more cost effective.

Flex (formerly Flextronics) operates one of the leading experiments in this field; it recently opened a factory in downtown San Francisco, a few blocks from City Hall and in proximity to mass transit. Flex is the second largest contract manufacturer in the world, behind only Taiwan's Foxconn. They are known for building industrial and consumer electronics products at scale in factories from China to the U.S. and Mexico. Flex is the very definition of a large, global, outsourced electronics manufacturer.

Steven Heintz is VP of Engineering and General Manager of the Flex Innovation Lab. As we toured the facility, Steve told us:

"We set up this facility to help product teams design for manufacturing and to manufacture your first few hundred units right here in San Francisco. You'll typically come here after you've gone through the TechShop phase. When you come through the doors here, typically you have a working prototype, but it might be some circuit boards duct taped together with some batteries and some sensors, and you've got a beautiful picture of what you want the product to look like.

"So here we're working with you to figure out, are your parts going to be injection molded, cast, or machined? In each case, how [does] the design needs to be changed? How [are} your circuit boards going to be affixed inside? How is the whole unit going to be assembled in a repeatable way? What test or inspection processes are going to be required for repeatable quality?"

On one side of the facility is fabrication, where products are designed and one-off prototypes made. On the other is engineer-assisted assembly lines. Here product developers work together with assembly techs drafting assembly instructions, creating the test fixtures and assembly jigs, working through the manufacturing process all the way to final test and pack out.

Taking the Risk out of Manufacturing

Says Heintz, "[In our San Francisco facility], we'll build the first dozen to several hundred units here for a customer. We'll get the process refined, so that then it can scale to a different geography—likely a higher volume site in the U.S., Mexico or China. But by then the risk is reduced as early small builds of products can be field-tested, and design improvements can be made before scaling to mass production."

Heintz spoke of the portfolio of products Flex was manufacturing in San Francisco.

"We mostly do consumer electronics here, so we do a lot of cameras, we do drones, we do a lot of wearables, because wearables are really hard to prototype using old-fashioned techniques like 3D printing. Think about how you would prototype a rubber membrane over the top of a flexible piece of electronics. You can't do that the old-fashioned way where you order some plastics over here, order your PCBs over there, and get your metal parts bolted all together, clamshell-style. We find that you actually have to be iterating and over molding, using the same techniques that [are] going to be used in production, before you even know if you have a viable product."

The old model was that a startup would have to make a fairly high volume commitment to a factory site somewhere in the world, committing to tens of thousands of units, before they'd even get the opportunity to make that first real working unit, and then stuck and held to those volumes.

A key small-batch manufacturing approach used by the Flex facility in San Francisco is low-cost 3D printed injection molding. Flex is now able to print a early low-cost injection mold overnight using plastic and start shooting the actual finished parts the next day. These molds are temporary in nature; you use them to yield only a few dozen finished parts, but they are more in line with the true finished product. These injection-molded parts contain all the nuances that will be needed for greater volume production—channels for adhesive, living hinges, even antennas molded into the plastics. They are also made with the actual resins, and colorants—demonstrating the material properties that the mass manufactured product will have.

Traditional injection molds take a couple months to produce. Now a manufacturer can build eight or ten iterations in the time it would take to get one injection mold.

"Instead of firing out email questions somewhere far away saying, 'Hey, can we do this or can we do that?' and getting a yes or no answer back, your product team is going to be able to call BS on that sometimes and say, 'You know what? We tried a different hold time or a different temperature or a different resin mixture or a different colorant blend, and we were able to get these results.' You're able to get tighter, less compromised products this way.

"After the temporary plastic mold is refined, an aluminum mold is machined. This may cost $6K-$10K, still far less than a steel injection mold. That mold can build a few hundred to few thousand early products, all still in San Francisco."

REMOVING CONSTRAINTS AND BUILDING LESS COMPROMISE INTO PRODUCTS

"[The U.S.] went through this phase where everybody thought all of manufacturing was headed somewhere else, and we lost a lot of our ability to understand, iterate, and innovate on the manufacturing process—engineers on our product teams lost some ability to push back. As a side effect, we also lost a lot of our ability to make tighter, less compromised products, because we took whatever the capabilities of whatever factory somewhere far away... happened to have available.

They import the actual resins that might eventually be used in Mexico or China, to encourage smooth scaling should the product need to move overseas for larger batch sizes.

"I see more product teams using new production and fabrication technology to place smaller bets. We're gonna release a limited-edition this, and we're only going to make 20,000 units but we're gonna try these three materials or technology/feature combinations. Another variant might be kind of experimental, so we want to try 10,000 of these. And so now you got a wider diversity of your product mix and SKUs. That becomes hard to do in an overseas kind of model, because it's now become a series of small batch runs."

Historically the venture capital community coached and financed startups to go to China and risk scaling production there. Today VCs are among Flex San Francisco factory's most attentive customers.

"We court the venture community probably more than anything else. A couple of years ago if a VC was going to give a local hardware startup up $10M, that hardware startup would find the cheapest manufacturer in China. They'd blow half their money on steel tools and NREs (non-recurring engineering). And they get their one shot at seeing what the product would look like and you saw so many products that never came to market and they blamed manufacturability issues. Now the VCs coach their start ups to spend a little more per unit and iterate on a product locally so we can all come over, we can all visit, we can all see what's going on, we can catch things earlier, we can get something to market earlier, and we can fail faster and we can manage that risk."

When we began work on the Maker City book, our biggest question was about what kind of industries were most suited for urban manufacturing. Surely, we thought, there were some industries like consumer electronics where the volumes, the components, the expertise, and the tooling costs all favored manufacturing offshore. Today, we think differently on this topic. Flex has created a kind of virtual factory that starts in San Francisco and spans to other plants in Mexico and China. Within this virtual factory, products can start out in the San Francisco-based Innovation Lab and move elsewhere within the Flex network only when they are ready to hit the next production target, and not before. Some products evolve so quickly or replenish in such small batches that it makes sense to avoid the complexity of "mass" manufacturing altogether.

MANUFACTURING START-UPS HEAD FOR THE CLOUD— AND YOUR CITY

BY MARK MUNRO

Senior Fellow, Brookings Institution

It's old news that an explosion of Internet-based tools and services—think Git Hub and Amazon Web Services—has made it easier and cheaper for entrepreneurs to transform their ideas into finished software products. One result: The urban start-up ferment now energizing city innovation zones from Nashville, Tennessee to Boulder, Colorado has tilted heavily toward software. Actually making things—and manufacturing them—has remained a more avant garde pursuit for the Maker Movement.

Yet now that's changing, with potentially big implications for industry—and cities. Thanks to the rise of an array of new tools, facilities, and services on the hardware side, a new "hardware renaissance" has begun to spread from Silicon Valley out across the U.S.

This renaissance builds on the spread of hobbyist "Makerspaces" and rapid prototyping studios equipped with 3D printers enabled by design software. And it has the feel of an insurgency. But in the last year a number of developments have put manufacturing startup activity on a faster, more commercial, and big-time track.

Entrepreneur Mark Hatch has opened eight TechShop Makerspaces in U.S. tech metros, such as the San Francisco Bay Area; Austin, Texas; Pittsburgh; and Washington, DC. These urban centers function as open-access, community-based workshops and prototyping studios.

3D printing and the crashing prices of microchips, sensors, and other components now make it thinkable for a small company to design sophisticated devices at reasonable cost, which Kickstarter and other crowdfunding sources to enable initial finance.

And now, a variety of specialized incubators and contract manufacturers available through the cloud of remote computing and other resources are making it easier for hardware dreamers and startups to obtain advice, support, and access to support services, including outsourced manufacturing. **As related last year by the Wall Street Journal's Chris Mims**, companies like PCH International and Dragon Innovation are now available to manage contract manufacturing and otherwise "make manufacturing feel easy" to entrepreneurs or small companies. Likewise, Andy Rubin, the creator of the Android mobile operating system, announced in April that his new company Playground Global LLC will serve as a sort of incubator "studio" where entrepreneurs and small firms can focus on building new gadgets while Playground takes care of physical-world challenges: engineering, manufacturing, scale-up financing, supply-chain management, and distribution.

WSJ | New Means of Production
mcbook.me/2csSjSm

Put it all together and it's pretty clear that a confluence of Maker craft-work using new tools, software-informed entrepreneurship, and commercial enterprise is working out the outlines of a whole new way of doing small-batch manufacturing of consequence that takes full advantage of such emergent themes as crowd funding, "manufacturing as a service," franchised maker-spaces, innovation labs, and manufacturing-oriented incubators.

The upshot: A suite of tools and supports like the ones that have fostered the software boom are now becoming available for hardware manufacturing.

Which opens up possibilities. For his part, Mims imagines an age in which "new products—actual, physical products—will go from idea to store shelves in a matter of months." Such an era could be beneficial for the U.S., given the nation's advantages in creativity and cloud-based business organization, even if much of the resulting new production could well occur offshore (though aspects of the "manufacturing as a service model" no longer ordain that!). What's more, the new ferment is clearly bringing about a fascinating convergence of Maker Movement craftsmanship and corporate reinvention.

But beyond that, such a surge could be transformative for cities. Currently, urban startup communities remain heavily virtual—all about software ideas and consumer-Internet ventures. That leaves urban economies narrower than they might be, and it likely forecloses on opportunities, since innovation is entwined with manufacturing. By contrast, the emergence of new cloud-enabled, incubator-supported manufacturing startups could widen the aperture. New synergies and new opportunities will be possible if physical-world inventors and entrepreneurs gain traction alongside virtual ones. Likewise, manufacturing enterprises could flourish without needing large exurban spaces. Ultimately, cities and their **innovation districts** will benefit if they can channel more of the hardware-oriented tinkering and entrepreneurship that launched Silicon Valley in the first place.

In short, the new tools and supports for hardware startups could be really important for cities. If hardware startups become more like software ones, and the U.S. can keep more of its high-value manufacturing onshore, U.S. cities and metropolitan areas will be able to bolster technology development and ground it here, to the benefit of America's cities.

→ Mark Muro is a senior fellow at the Brookings Institution where he leads work on advanced industries. Portions of this commentary are reprinted from The Avenue, a blog at **Brookings**.

Brookings | Innovation Districts
mcbook.me/2cffzQv

Brookings | Avenue Blog
mcbook.me/2cqyFTM

IMPLICATIONS FOR CITIES

To encourage manufacturing in your city, consider the following:

 Connect up with the Urban Manufacturing Alliance (UMA) to learn additional best practices and how they can apply in your Maker City. Visit www.urbanmfg.org

Compete for funding from the SBA's (Small Business Administration) Growth Accelerator Fund[8]. The SBA has set aside $3.95M for grants of $50,000 for small businesses that directly support entrepreneurship and small-batch manufacturing. Makers seeking to do small production runs of 500–10,000 units locally in the U.S. tell us they have a particularly challenging time finding financing. Make them a priority.

Use real estate strategically. Deborah Acosta, the Chief Innovation Officer for the City of San Leandro, told us that in her Maker City they give manufacturing concerns "first call" on space over distribution centers. Manufacturing creates more jobs and economic activity versus a distribution center.

Consider subsidizing adaptive reuse projects that will bring manufacturing back to your city. Ideal targets are former shipyards, factories, and industrial space. Legacy space that is larger but still has character is particularly attractive to Makers, but it needs to be cut up into smaller spaces. Look for proposals from real estate developers who are willing to provide rent stabilization to Makers and/or manufacturing concerns. The Pratt Center believes that the best model for adaptive re-use is to put a mission-oriented nonprofit in charge of the effort. To learn more, check out this **toolkit**[9].

Think regionally. Link up local Makers with providers of Manufacturing-as-a-Service. The goal here is to give Makers in your city the same access to manufacturing and supply chains that big firms take for granted. Proximity is important to reduce cycle time but so is access to advanced equipment and precision engineering techniques. This type of clustering of manufacturing capacity and supply chain is what once made the Rust Belt great and will make it great again.

Look at ways to encourage advanced manufacturing through regulatory relief, zoning relief, and tax abatement. Zoning relief is a city matter we discuss more in Chapter 7 on Real Estate; tax abatement typically comes from working with the State to provide relief. By thinking regionally (as discussed above), Maker Cities can make tax relief more palatable at a state level.

Encourage small manufacturing firms to take root. Ask the EDC (Economic Development Corporation) or mayor's office to do a survey to identify Makers who want to produce their products locally, but don't believe they have the infrastructure they need for startup manufacturing.

Invest in the right kinds of manufacturing. Manufacturing firms that have the most potential for growth are those that use advanced manufacturing prowess or are R&D intensive, according to the consulting firm **McKinsey**[10] writing in 2012. Look for firms that can source materials locally to replace items previously purchased from overseas. Also, look for firms that can create a product to export to the rest of the world, due to uniqueness in underlying technology or process.

Encourage manufacturers in your city to get together and share best practices. Manufacturers have a lot they can learn from each other. Bring manufacturers together in small round table settings; curate the meetings with care, so as to make sure that direct competitors are not in the room. This encourages the manufacturers to share process know how in a greater level of detail. We believe this recommendation is key to encouraging more startups to take advantage of custom manufacturing at point of sale (LYF example), crowdsourcing (FirstBuild example), and precision manufacturing of components (Portland aerospace example).

Consider a hardware accelerator model. Product design and manufacturing continue to be ongoing challenges for Makers. A hardware accelerator can help, by providing experienced people to coach and mentor Makers through the process. Some hardware accelerators also provide venture financing, access to a well-equipped Makerspace, as well as co-working space. Hardware accelerators are a relatively new concept and do not exist in every city. Examples include: AlphaLab Gear (Pittsburgh, Pennsylvania); Lemnos Labs and PCH Highway1 (San Francisco, California); and Playground (Palo Alto, California).

Focus economic development officers on encouraging small manufacturing firms to take root in your city. The old model of economic development in our cities was to focus efforts on landing the "big fish." The data tells us that **most manufacturing firms start out small**[11] (approximately 70 percent of urban manufacturing firms employ fewer than 20 people) and stay that way. The economic value when aggregating all these small firms together can be large and have a transformative impact on your city, as seen in the examples from Fremont and San Leandro.

Activate research universities as partners. As noted above, American cities are particularly good at manufacturing products with strong R&D inputs.

Work with the investment community in your city to create new forms of financing. Many banks have community development goals they must meet. Encourage them to invest their dollars in ways that support manufacturing in your Maker City. Banks may be ideal partners to provide initial funding for a hardware accelerator in your city. Additionally, partnerships with Mastercard, Visa, and American Express may be helpful. Mastercard has a particular focus on factoring, which is when a financial services firm lends money to a manufacturing firm based on the future value of its inventory. Of course, today's manufacturing firms do not hold a lot of inventory, making factoring less relevant as a source of financing. But we fully expect a new class of financial products to be invented, focusing on the new and evolving needs of manufacturing firms inside our Maker Cities.

RESOURCE

QUICK WIN

IMPLEMENTATION ADVICE

POLICY

BIG IDEA

URBAN MANUFACTURING ALLIANCE

Tying Together Best Practices Across The Country For New Maker Communities

BY KIM-MAI CUTLER
Reporter, Tech Crunch

When U.S. manufacturing employment peaked in the late 1970s, American cities had eroded, with their economic fortunes reaching a nadir. New York City infamously almost filed for bankruptcy and 97 percent of the buildings in certain Bronx census tracts had burned to the ground through suspected arson-triggered fires.

Suburban decentralization attracted jobs from the urban core, while automation and globalization ensured that many manufacturing jobs disappeared altogether. What was left behind were former industrial spaces that quickly became absorbed by a highly polarized knowledge-and-services economy with white-collar workers and lower-wage service workers.

Cities rebounded, but manufacturing employment didn't—at least not until the end of the last recession. More intriguingly, manufacturing's real output and its real output per capita in the U.S. have increased while employment has dropped over the last forty years. Out of all the developed countries that had strong industrial bases in the twentieth century, the United States is the one that has seen its manufacturing sector fall faster and farther in terms of its share of GDP over the last half-century when compared to nations like Germany and Japan.

Today, in the face of both real estate pressures and globalization, American cities have choices to make about whether to preserve and enhance their remnant manufacturing bases and even grow nascent Maker communities.

These choices are not accidental. They are deliberate. That's part of the reason that the **Urban Manufacturing Alliance (UMA)** came into existence five years ago.

"There was a realization that we would be stronger in numbers if we wanted to see any global initiatives or any policies supporting local manufacturing," said Lee Wellington, who is the Executive Director of the Urban Manufacturing Alliance. "There are so many cities doing disparate research projects and the methodology is so different in each one. What we want to become is a clearinghouse of all of these ideas in the Maker movement."

Kate Sofis, along with Alliance co-founder Adam Friedman honored onstage by President Bill Clinton for their commitment to grow US manufacturing jobs. Photo courtesy of SFMade.

UMA
urbanmfg.org

CASE STUDY
URBAN MANUFACTURING ALLIANCE

163

Wellington came to the nonprofit after working in the New York City's mayor's office and seeing how urban policy affected the trajectory of the Brooklyn Navy Yards, where her great-grandfather worked about a century ago shortly after immigrating to the United States from Russia.

Like a think-tank and advocacy organization, the alliance, which has more than 100 members across the country, pulls best practices in workforce development, equity, real estate, and local branding together.

It has to speak to manufacturing's past and its future, both representing communities that echo the industry's historically blue collar roots as well as newer strands in the Maker movement, which make small-batch or artisanal goods at higher price points but will eventually need to scale their headcount and footprints.

"The manufacturing industry has been a stabilizer for so many families in the United States. But it can also be a very dirty word," Wellington said. "People, like my family who worked in manufacturing never wanted their kids to work in manufacturing. So we have to shift the perception on a couple different levels."

The American manufacturing sector has changed dramatically over the last generation. For one, employment began rebounding in 2010. Instead of large firms, smaller ones are forming and they're located more closely to end consumers. Large, vertically-integrated factories are giving way to smaller, more flexible firms that can rapidly respond to changes in consumer tastes or technological breakthroughs.

While UMA is mostly focused on local and municipal efforts, because that's where policy most deeply affects Makers and manufacturers, it is trying to branch into higher federal level policy. The organization is trying to influence a federal New Markets Tax Credit Program, created back in 2000, to become an additional capital source for cities looking to build new industrial or Makerspaces.

To re-imagine what Making and manufacturing could look like going forward, UMA has knit together member communities from Detroit and Cincinnati to San Francisco and Seattle.

They do this through about a half-dozen communities of practice which include:

→ Workforce Development

→ Equity

→ Land Use Policy and Real Estate Development

→ Local and Regional Branding

→ Sourcing and Supply Chains

→ Manufacturing Policy

We're going to walk through a few of these with case examples from UMA member cities:

1. Local Branding and Marketing

With the complexity and opacity of modern, globalized supply-chains, provenance and local pride are two of the things that urban manufacturing and Maker communities can tap into to connect with and reach consumers.

Founding UMA member SFMade was one of the earliest pioneers in creating a local branding effort that put together everything from a common logo to retail partnerships. When they began, they started with 12 founding members that ranged from the city's longest-standing manufacturer, McRoskey Mattress Company, to newer companies like Ritual Coffee Roasters and DodoCase. They wanted to make sure that everyone from legacy brands to startups could feel included.

To promote local Makers, they started an SFMade week and special days where member retailers would donate 10 percent of their sales to the organization. They also held online auctions and factory tours, and created a retailer map sponsored by Levi Strauss & Co., a classic San Francisco brand from the original Gold Rush era. They also worked with the city's travel and convention authority along with hotel chains to promote the map to visitors.

To figure out who could qualify for the branding program, they set out very specific criteria. Businesses had to be headquartered in San Francisco, manufacture at least one product in the city, have a local workforce with at least one full-time employee, and create consumer-facing products that could either be sold in retail stores or online. If they fit the bill, they could get the right to use the SFMade logo, be listed in an online directory, and cut sales deals through SFMade events and retail relationships.

To make the brand sustainable, SFMade worked on a voluntary pay-what-you-will basis. From that, they began earning about $60,000 every year from about 400 member manufacturers. On top of that, they began running educational programming with two-hour workshops and one-hour advising consultations, which brought in an extra $20,000 in revenue.

Across the country, Made in NYC, a local branding effort that is more than a decade old and was founded in the wake of the September 11 attacks, has stepped up with a larger-scale advertising campaign through a $750,000 grant from the New York City Council. More than 1,300 companies are now part of the effort, which has had recent campaigns under the slogans "Dreams, Jobs and ____," and, "Made Here in NYC."

Back on the West Coast, Portland's local branding program follows a similar model, with a membership-driven structure that sustains itself through professional services that help Makers with business strategy and marketing. From $100 to more than $1,000 year, local Makers can be a part of the brand. On top of that, Portland Made offers by-the-hour services in photography, web design and real estate match-making. They'll also help connect Makers and designers with factories that can do small production runs.

"In Portland, we have a pretty long history of manufacturing. We had a lot of shipbuilders. We had iron workers and lumber and wood-oriented products," said Kelley Roy, who started Portland Made after opening a large Makerspace on the eastern side of the city called ADX. "But we're also a hotbed for creativity and entrepreneurialism. Portland has always had this this kind of do-it-yourself, very independent spirit."

The toughest issues facing Roy and Portland Made are how to help Makers in that mid-stage—companies that are five to ten years old and need to have a proactive sales strategy but still don't have enough of a budget to put capital toward those needs. Roy partners with Made Here PDX to carry dozens of locally-made products.

2. Accelerators for Early-Stage Maker Startups

In the center of the country, Cincinnati Made takes a slightly different approach with a hands-on accelerator called First Batch that is much more intensive than the SFMade and Portland Made membership models.

During the program, the nonprofit provides four months of free rent at a Makerspace along with up to $10K in assistance for manufacturing and production and connections to mentors and advisers from the Cincinnati manufacturing community.

What Founder Matt Anthony discovered after graduating in an industrial design program was that a lot of the city's young talent would leave despite Cincinnati's strength in consumer packaged goods with major companies like Procter & Gamble.

"We have one of the top industrial design schools in the country. For most people, the end goal is leaving and going to Seattle or SF," he said. "We talk about brain drain way too much, even though we have all this heritage manufacturing in the city."

First Batch picks around eight companies with each cycle based on the strength of the team and idea, and their commitment to doing business in Southwest Ohio. Anthony said they've had a few breakout hits, including a beard oil and a high-end razor brand that has generated around $200,000 in commitments on Kickstarter.

"We never had a distinct crash like Detroit. When you think about these Rust Belt cities, all of them have to reinvent themselves on some level. Pittsburgh had to basically rebrand from steel and they came out of it with robotics," Anthony said. "But Cincinnati has always been about consumer products and we never saw a crash from people not buying soap or detergent. We can be adverse to change but there are always little threads that you can pull out of our story. We do have a lot of resources around small-batch manufacturing."

3. Land Use in Hot and Cold Real Estate Markets

While we do have a land-use and real estate section in this book, the main dilemma that UMA faces and advises has to do with the hot-and-cold nature of the real estate market.

In certain coastal cities like New York City and San Francisco, the dilemma is about how to preserve industrial space when there are extraordinary market pressures to convert them to more lucrative uses such as condos or office space. For new construction, the policy suggestion is to have cities build in incentives for mixed-use development that puts Makerspace in alongside residential or commercial space. To build understanding with the broader public, UMA recommends that Makers open their spaces to the community through brewery tours or open houses and events so that voters understand the value of Maker or industrially-zoned land.

But in less highly demanded areas, the problem is that rents are too low to cover basic maintenance or upkeep for properties.

In both cases, UMA recommends establishing or working with nonprofit development corporations, a strategy that has been used for decades in affordable housing development. Nonprofit developers can take a much longer time horizon and don't need to sell off properties at certain points in time or parts of the economic cycle for their investors. They can also re-invest their profits into acquiring or rehabilitating more properties to build out a portfolio. Their funding sources tend to be much more diverse and they sometimes need years to establish the right capital stack to take on a project. The best-known projects tend to be hybrid public-private corporations like the Brooklyn Navy Yard.

One project to take a look at is Ruckus, a new Makerspace opening in Indianapolis in fall 2016 that was created by a nonprofit developer, Riley Area Development Corporation. Although Riley has been around since the late 1970s in the city, they partnered with the city of Indianapolis in taking a grant that covers 20 percent of the total project cost. They're building a Makerspace in a 540,000 square foot former factory, which sits at the gateway of a 500-acre industrial redevelopment district. In other parts of the factory will be space for food retailers and offices and artists studios.

RUCKUS
Makerspace—Riley Area
Development Corporation

On top of that, Indianapolis has undergone its very first major re-zoning in decades with new policies on the books that are intended to attract artisan manufacturers and food producers into the city's vacant buildings.

"Indianapolis has lost over 25,000 manufacturing jobs. There have been really huge shutdowns of automotive plants and now there's so much more brownfield space. It's a huge loss," said Eric Strickland, who is Riley's executive director. "The key is not fighting politics but building political will with a real understanding of the real estate market so that you can build political interest and not compete against every other economic development idea."

Even with help from the city, it has been tough to raise the rest of the funding for the space, Strickland said. Potential financiers perceive the Makerspace as risky and so they're charging a higher interest rate to compensate for that. Riley is still trying to close another $3 million.

"It's still very difficult. The market does not always see the economic benefits. We're somewhere between what you'd see at a WeWork or co-working space to a full Makerspace," he said. Ruckus will have everything from plain old desks to extra equipment like laser cutters to drill presses that members can access on premium price tiers.

4. New Training Strategies for Youth

Although the workforce development component of the alliance's toolkit is still just getting started, the nonprofit wants to take a special focus on bringing young people into the industry. They're starting with a series of partnerships in the Philadelphia public school system and are taking a closer look at some programs in Chicago and San Diego.

In Chicago, Dan Swinney is a labor activist who spent 13 years as a machinist and is now behind a program in a low-income, historically African American area of the city that's trying to train teenagers in getting into manufacturing jobs. He founded a school called Austin Polytechnical Academy, that recently won $2.7 million dollar grant from the U.S. Department of Labor to build out its Youth CareerConnect program. That school led to the creation of a broader program called Manufacturing Connect, that serves students across three schools in the area.

"I was studying the skills gap problem," Swinney said. "We found that starting in the late 1960s, the manufacturing industry began to change. Back then, the whole notion of manufacturing was perceived, for both the right and wrong reasons, as dangerous and dirty work. Now, we've shifted to higher value added and more complex work. But what's degraded are the parts of our education system that are linked to manufacturing."

Swinney began looking at more vocational-focused education systems in other countries like Germany to see what he could borrow and bring back to Chicago. He idealized structures like Mondragon, a federation of worker co-operatives in Spain that has more than 74,000 members and 11 billion euro in revenue.

"In the U.S., it's broadly assumed that only the private sector with its principle values of private accumulation of wealth is capable of leading the development of the manufacturing sector," he said.

But Mondragon, based in the Basque region of Spain, provides an alternative view of what labor and production can be. Developed toward the end of World War II by a young Spanish priest,

Mondragon became a collection of cooperatives that is governed through a combination of technical experts and elected councils. The majority of profits are then re-invested back into the businesses every year, there is an unemployment assistance fund that helps kick in during recessions, and the pay ratio between managers and the lowest-compensated workers is restricted.

In the way that José María Arizmendiarrieta started Mondragon as a vocational training school, Swinney proposed the idea of creating a high school that could meet the standards of the local manufacturing industry.

As part of Manufacturing Connect, freshmen and sophomores will go on tours to see what modern manufacturing jobs look like. They then go on to take skills classes like learning how to use CNC machines later in their junior year. Paired with all of this are internships and job shadowing during spring breaks. Throughout, students take three to four years of pre-engineering courses in a system called Project Lead The Way, where engineers create the curriculum and courses discuss the issues of design, materials, and measurement

Since the program started, about 47 percent of its graduating students have participated in a paid manufacturing work experience, and at least half have earned at least one metalworking credential from the National Institute of Metal Working. Of the 360 students that graduated from Austin Polytech from 2011 to 2015, Swinney estimates that 150 of those students were active in one way or another in the Manufacturing Connect program. Of those, around 50 have pursued careers in manufacturing. Around 38 pursued full-time jobs in manufacturing with an average retention rate of one year making between $20K and $75K a year. The rest—10 to 12—pursued their manufacturing interests in college through studying engineering.

"Our competitive advantage—the U.S.'s competitive advantage—is in advanced manufacturing. We are still within range of keeping that. But if we don't revamp our education system, we will lose that advantage," Swinney said. "It has to do with demographics and the loss of the baby boomer workforce. So for me, the next ten years is really critical. If we lose our competitive advantage, we will become marginalized in global society and you can feel that already happening with polarization and income inequality. So I have this sense of urgency."

5. Ensuring That New Maker Communities are Equitable

The last major piece that the Urban Manufacturing Alliance is starting to examine is inclusion and equity. Wellington says a full toolkit is going to be out shortly, with studies on New York, San Jose, Portland, and Indianapolis. This section, unfortunately, is less fleshed out because the equity and inclusion toolkit is still not finished at this point in time.

However, one of the cities that's a part of the alliance's very nascent equity efforts is San Jose, which is the large, often over-looked city that sits at the bottom of the San Francisco Bay.

"San Jose has a really unique story that's valuable and needs to be more of a part of the Bay Area narrative," said Michelle Thong, who is a business development officer at San Jose's Office of Economic Development.

The city, which is the most populous in the region, doesn't get the same media attention as San Francisco, which has sucked much of the tech workforce and its sources of capital northward from the industry's historical core in Bay Area suburbs like Palo Alto, Menlo Park, and Mountain View. Nor is it getting the same cultural attention as Oakland, which has received an influx of San Francisco's creative workers and industries after being priced out of the Western half of the Bay.

Yet San Jose's industry in many ways is more diverse than that of the broader tech industry, which has seen widespread criticism for its lack of women and underrepresented minorities. Other East Bay cities like San Leandro and Fremont have long histories with the manufacturing industry, most famously with Fremont's General Motors plant, which started in the mid-20th century, but then evolved into the Toyota joint venture NUMMI and is now a major Tesla manufacturing location.

Thong said San Jose's advanced manufacturing sector is about 23 percent Latino. African Americans are still underrepresented at 1.8 percent of the manufacturing workforce, but they also only make up 3.5 percent of the city's population.

"Many urban planners by training have come to this consensus around this shiny, happy new urbanist vision that has transit and amenities," Thong said. "But there's also this bifurcation. Where are the middle-income jobs? Where are the industrial lands and uses?"

The alliance is encouraging Maker programs to start looking at census tract data to check workforce placements or analyzing whether venture capital and investment is going toward underrepresented minorities and women in local communities. Cities that want to develop effective policy need to also start examining the skills gap to see how much on-the-job technical training is required.

They could ensure that Maker or manufacturing spaces pay a living wage or not exclude job applicants with criminal backgrounds, which penalize communities that are already disproportionately swept into the criminal justice system.

→   Kim-Mai Cutler is a reporter at TechCrunch.

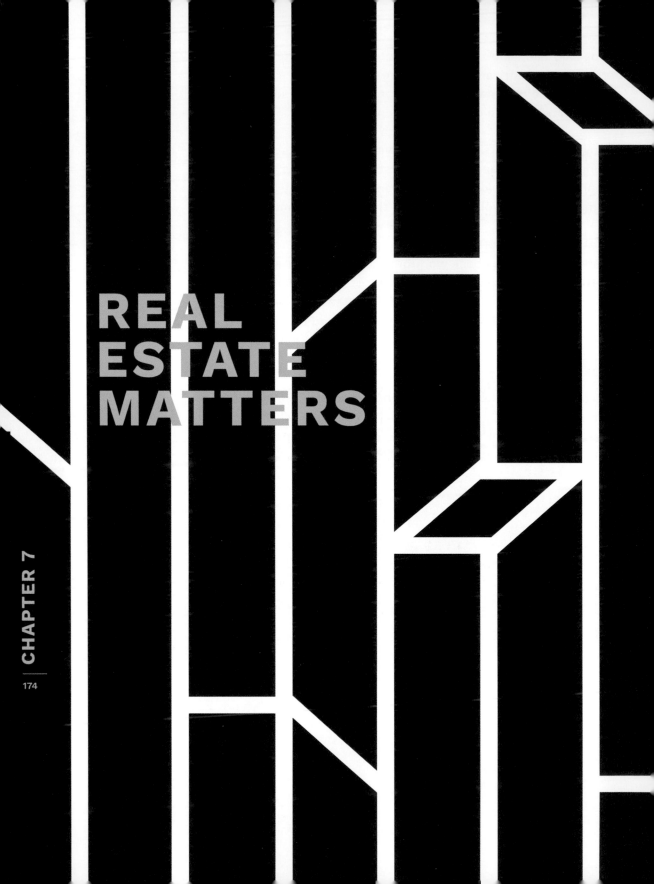

REAL
ESTATE
MATTERS

MAKING SPACE FOR MAKERS

Real Estate Connects the Dots between Creativity and Value Creation in the Maker City

The form our cities take has always expressed what we value most in society. Cities of the middle ages gave us great cathedrals and castles, prior to the age of enlightenment, when God and king were central to our universe. In the twentieth century, cities built out cathedrals to commerce and production in the form of skyscrapers and industrial plants.

Co-working. Innovation centers. Makerspaces. These forms of real estate are not only relatively new, they are startlingly different than past expressions of commercial real estate. Each is designed to break down barriers between people, build a sense of community, embrace technology, and encourage experimentation and the creation of new forms of economic value inside our cities.

What kind of space will best serve the Maker and innovation economy is suddenly an important topic in urban planning. Major companies are abandoning the suburbs in favor of cities. GE is moving from suburban Connecticut to urban Boston. The center of gravity of Silicon Valley has shifted north to San Francisco and Oakland. Portland is turning its considerable strengths in technology and urban manufacturing into unprecedented growth.

IN THIS CHAPTER, WE LOOK AT ZONING, LAND USE POLICIES, AND POLICY HACKS THAT CAN BE USED TO MAKE ROOM FOR MAKERS INSIDE AMERICA'S CITIES.

Real estate is of interest to the Maker City for four main reasons:

→ **Makerspaces.** Makerspaces. Makerspaces can range from 5 thousand square feet to 150–200 thousand square feet. There is no average or "best size" for Makerspaces. Larger spaces are prized both because they support the widest variety of tools and functions and because they can be divided into many smaller spaces, as needed to support small workgroups, multiple classrooms and workshops, even co-working. Increasingly, we are seeing Makerspaces placed in innovation districts, as discussed below.

→ **Urban manufacturing.** Americans are showing a renewed preference for buying locally made goods in a wide variety of categories. Astute retailers like West Elm and Etsy are getting in the act, curating goods locally and selling them through both virtual and real storefronts. The sum of artisanal manufacturing plus advanced manufacturing (as discussed in Chapter 6) adds up to a big need for space. Adaptive reuse of older shipyards and factories is the most promising approach.

→ **Innovation centers.** City, business, and nonprofit leaders often seek guidance from experience design firms when commissioning creative space. When cities devote an entire neighborhood to innovation it's called an "innovation district" or sometimes an "innovation zone." (We prefer the first term.) When a single company or nonprofit builds out a facility for innovation it is called an "innovation center" or "innovation lab," depending on its size. This can get confusing. Adding to this confusion is that many cities in the U.S. are pursuing ways to turn whole areas of the city into living laboratories for urban innovation in order to test the impact of sensors and other "smart city" technologies.

→ **Live/Work housing.** Makers require support in the form of affordable housing, ideally with adjacent studio space.

In this chapter, we'll showcase what is possible when underutilized real estate is put to use for startups, artists, Makers, and manufacturers through three case studies: **Brooklyn Navy Yard, Manufacture NY, and the Chicago 1871 Project.**

We'll follow these case studies with contributions from three experts in real estate:

→ **Adam Friedman, Executive Director, Pratt Center for Community Development.** Its mission is to provide technical assistance, research and policy on land use, economic development, and community organizing in New York's low-income communities. The Center is a founding member of the Urban Manufacturing Alliance, which does research and work to strengthen the urban manufacturing sector on a national level.

→ **Kim-Mai Cutler, Reporter, TechCrunch.** Kim-Mai has written extensively about the affordable housing crisis in San Francisco. Here she discusses policy hacks that cities like New York and San Francisco can use to reserve space for Makers.

→ **Heather King, Managing Director, Eight Inc.** Eight Inc. is the experience design firm behind the Apple Stores, which have proven to be an inviting destination for both city residents, commuters, and visitors. Most relevant to us in this book is what Eight Inc. does when working with cities to creating innovation centers like Future Cities Catapult (FCC), an Innovation Lab based in London. We asked Heather King from Eight Inc. to tell us how they create innovative spaces that capture people's imagination while getting diverse stakeholders to work together.

Together, these experts provide a nuanced view of how cities can make space available to Makers and then design it in such a way that people come together to do their best, most imaginative work, the kind of work that adds economic value to the city. No space exemplifies this more than the Brooklyn Navy Yard.

THE BROOKLYN NAVY YARD

Learning What Works

David Ehrenberg has served as Executive Director of the Brooklyn Navy Yard Development Corporation (BNYDC) since 2013. For 20 years the economic development approach of choice to revitalize the Brooklyn Navy Yard, or any other former factory or shipyard for that matter, was to bring back a single big employer or a factory, preferably a shipyard or an auto plant. Efforts to do so repeatedly failed. It was only through the involvement of a mission-driven nonprofit, the BNYDC, that the Brooklyn Navy Yard started to make a comeback as a community housing many smaller businesses made up of Makers, artists, artisans, and manufacturing concerns.

Today over 330 business and 7,000 people work out of the Brooklyn Navy Yard in 3.5 million square feet growing to 5.0 million square feet by 2018.

According to David Ehrenberg:
"[W]hat we've been able to curate here, because we have such a large footprint, is a wonderful chaotic mix of tenants doing sand-blasting, doing robotics work, doing just pure arts, doing pure architectural design. And they all find each other and collaborate, and create this kind of gestalt of the Navy Yard that is extraordi-narily hard to find anywhere else in New York.

"[I]t was kind of a natural organic process where, because we were on-site managers and heard from tenants all the time that what they needed was a five thousand-square foot or a two thousand-square-foot unit. And they didn't need that 60- or 100 thousand-square-foot unit. We were hoping, frankly, that we would be able to find a huge manufacturer who would take 200 thousand square feet. Who wouldn't want that, as a landlord? We were just hearing over and over and over again from tenants, 'No, what I really need is 500 square feet so I can produce my art,' or whatever it is.

"We've been 100 percent full for ten years. We've added very little new space here at the Yard, and we finally cracked that code here. Over the coming two years, we'll be adding almost two million square feet of space to the Yard, and then additional space over the coming five years."

Starting Small; Scaling Up

The true promise of a space like the Brooklyn Navy Yard is to allow smaller manufacturing firms to flourish side-by-side with larger companies like Crye Precision, producers of high tech body armor. Started by two graduates of Cooper Union with a background in engineering, Crye Precision has grown to be the one of the largest providers of high-tech body armor to the Department of Defense.

David Ehrenberg calls out Crye Precision as impossible to categorize:

"They are a design firm. They are a technology firm. They've got all kinds of IP that they can't tell us about because of national security concerns. But they're also hiring as many seamstresses and metal workers as they can to make the body armor and sew the camouflage."

The Brooklyn Navy Yard, in its role as a commercial landlord, sees a huge number of companies that have the potential to be the next Crye Precision. "These range from furniture manufacturers who are integrating technology into their furniture or just have a high tech means of production, to a company who's designing the next-generation of [electric] motorcycles."

For Cyre Precision, vertical integration makes sense. This is a company with expertise and know how in advanced material science as well as manufacturing. Advanced materials for textiles is of particular interest, as we'll see below in the case study on Manufacture NY.

The Impact of the Brooklyn Navy Yard

→ Economic output in 2011: $1.93B

→ Direct and indirect jobs: 10,350 direct and indirect jobs

→ Direct earnings: $392M in earnings

This economic activity in turn generated another $1.96B in earnings and an additional 15,500 jobs. (Source: **Pratt Center, 2013)**[1]

MANUFACTURE NY

Revitalizing the Garment Industry

Adam Friedman is the Executive Director, Pratt Center for Community Development. He talks about the garment district in New York City as "the original innovation district."

Today, much of what made the garment district a source of innovation has moved to Brooklyn in the form of a 150,000+ square-foot facility called Manufacture NY that serves as a national hub for the fashion and textile industries, focusing on design, production, training, and research.

"The idea is to bridge the gap between technology and the commercialization of products made of functional fabrics," said Fashion Institute of Technology President Dr. Joyce Brown.

Manufacture NY is part Revolutionary Fibers & Textile Manufacturing Innovation Institute, which in 2016 was funded through a $75M federal grant from the U.S. Department of Defense. The institute is creating facilities across the country that will help American industry develop commercial and military products with novel properties such as flame resistance, extremely lightweight, and embedded electronic sensors.

In addition to federal funding, the city awarded Manufacture NY $3.5M to build the 160,000 square-foot training center. The facility is expected to increase the number of highly skilled workers and create more manufacturing jobs in New York City. Also, it will boost the competitiveness of local businesses in the fashion and military-technology markets by helping people produce items like sweaters made of functional fabrics that regulate body temperature or parachutes with built-in radio transmitters.

"This is the next generation of gadgets," said Bob Bland, Chief Executive Officer of Manufacture New York. "Functionality will be embedded in the textiles themselves."

Real estate is a key part of the strategy to revitalize the fashion and textile industries, as Bob Bland explains:

"Fashion designers, students, Makers, manufacturers, showrooms —all of these groups need to be located together to have any hope of staying competitive. Cycle time matters a lot in fashion. Consumer taste is fickle. And fashion items cannot be copyrighted.

"By the time you take your sample to Asia—say to put it into production most cheaply—it is likely that the competition will have taken a similar design to a manufacturer closer to home.

"The businesses we work with need support over the long term, and it starts with general operations. Overhead is where people are getting killed right now, especially in New York City. Their overhead expenses include their rent, utilities, payroll, all of that. And so any change in their circumstances can be incredibly volatile, especially for contract manufacturers. They're typically working with 10–15 percent margins because that's the price that the customer expects. And there's no way to get out of that model without ballooning your margins and losing all of your customers.

"You can't afford for someone to only offer you a one-year lease. It's all about affordable, long-term space and a community platform that provides support in all the areas in which business owners don't know what they're doing or need help."

CHICAGO'S 1871 PROJECT

Think Small. Think Local.

It is easy to dismiss the Brooklyn Navy Yard and Manufacture NY as outliers. After all, what is possible in New York City is often not possible elsewhere in the country. But our research for this book uncovered a marked shift in the conventional wisdom on how best to use real estate as a development tool in the Maker City.

The normal plan when developing real estate designed to drive economic development is to lure a branch of a global manufacturer. But then you have an industry without roots, without a multiplier effect on the community.

"They're not using local accountants and local printers," says Susan Witt, Executive Director of the Schumaker Center for New Economics in Great Barrington, Massachusetts, which, since its inception in 1980, has been known for its close working relationship with the late Jane Jacobs. "It's through those roots that you get the economic multiplier effect of small businesses. And a branch or factory based elsewhere can leave as easily as it arrived."

Michael Shuman, Research and Public Policy Director of the Business Alliance for Local Living Economies, says research suggests that subsidies to attract and retain development are not effective at jumpstarting economies. One unpublished study he led recently looked at the three largest economic development programs in 15 states and found that fewer than ten percent of

Witt and Shulman
interview
mcbook.me/2cwl2mW

Companies involved devoted even a small majority of expenditures to local businesses; in most cases 90 percent of the money spent went out of state.

"The economic developers I speak to no longer even try to defend these subsidy strategies," Shuman says. "They've run out of excuses except for the fact that the politicians like them. Politicians get more mileage from one big deal that brings in 1,000 jobs than an entrepreneurship program that generates ten jobs in 100 local businesses. Even when the rhetoric has shifted to the importance of local in terms of where the money goes, it's still following an old and entirely discredited mode of economic development."

The "new normal" is a model of economic development that uses real estate to attract and support small and local product companies.

An example can be found in Chicago's Merchandise Mart. Built in the 1930s, the Merchandise Mart has gone through many boom and bust cycles. Today, it is home to 1871, an incubator and innovation hub which opened in 2012. Founder and CEO Howard Tullman says 1871 is the largest incubator in the country with over 500 startups in residence. Over 2,000 people each day work at 1871, which added 41 thousand square feet in 2016 to its original 50 thousand square feet for space.

Tullman's 1871 is more than just an incubator. It's a hub for innovation that not only attracts startups but also larger companies who want to be near it and lease space in the Merchandise Mart. Tullman says that the goal of 1871 is to create jobs in Chicago and keep the money raised by startups and the earning from their successes in Chicago. It has become a place that attracts people with innovative ideas as well as professionals with industry experience who seek to join a startup. It provides mentoring and education for startups, but Tullman believes that their most important function is "matchmaking," which helps startups build great teams and also helps startups find funding and corporate partners.

1871 could be seen as a new kind of lightweight business structure, consisting of many autonomous units that share a set of core services but also leverage each other in unexpected ways.

1871 thrives in the middle between startups who find it difficult to function in isolation and larger hierarchical companies who find it difficult to foster innovation. 1871 enables innovators to start small and collaborate with others while creating a larger ecosystem that drives innovation in the local economy.

ARE INNOVATION DISTRICTS THE ANSWER?

BY ADAM FRIEDMAN

Executive Director, Pratt Center

Cities across the United States are exploring how to capitalize on the entrepreneurial energy and talent being unleashed by the Maker movement. Planners and economic development practitioners are looking for ways to use this energy to generate new businesses and job growth, and to help revitalize legacy manufacturers. What are the land use challenges and strategies that cities should consider as they develop initiatives to build Maker cities?

One of the primary strategies toward this goal has been to create "innovation districts" that cluster universities, software and engineering companies, arts organizations, and other "creative" resources in proximity to manufacturers.

In this model, Makers play critical roles both in the origin of new products and in helping to translate the ideas from the "creatives" into the products for the manufacturers: they help generate the concept, they design and build the prototypes, they may market test the first small batches, and they refine the product before commercial scale production.

Towards Mixed Use Land Policies

A fundamental challenge arises out of the need to have a diversity of spaces at differing prices in relative proximity to each other. The manufacturers, the Makers, and the artists need cheap space. Designers, architects, and software and engineering companies can generally afford to pay significantly more for space. In a totally free market, over time real estate developers will decide to provide space for the higher-rent tenants. The lower-rent users who made the neighborhood both attractive for private investment and who advanced the public policy objectives of business and job growth will be priced out. The picture is even more complicated by the aesthetics of the industrial space and the emergence of "industrial chic." The attractiveness of mixed-use districts makes them unstable if property owners can easily convert from low-rent to high-rent uses. While property owners may oppose the restrictions that balance usage, such restrictions are essential to both the overall public and private value of the district. These low-rent uses are "the innovation commons" from which every property owner benefits but which no property owner wants to be responsible for providing.

Not only is preserving lower-cost space essential to maintaining the attractiveness of the higher-cost space, it is essential to achieving the type of broad-based economic recovery and gener-ating the new jobs needed to address today's unemployment and underemployment. A robust recovery requires that cities create not only jobs in the innovation economy (conceiving, designing, and making the prototypes) but also that they capture the next ripple of jobs as companies move past the initial innovation phase of their products' life cycles and into broader production for consumer markets.

There are two basic approaches that might be combined to support the diversity of spaces that is essential to innovation and growth. Zoning can be used to prescribe not only what uses are permitted in an area but a percentage of space in a building. Ownership or management of the production space can be vested in a nonprofit organization whose mission is to strengthen the manufacturing sector and to ensure not only the continued afford-ability of the production space but also provide services that support the companies and facilitate resident employment.

The Importance of Zoning

Long before the Maker movement, the original innovation district was NYC's Garment District.

For example, long before there was a Maker movement, the fashion industry inspired a similar type of Maker ethos. The Garment

Center in New York was, in some ways, the first innovation district. The Garment Center brought together a vibrant mix of fashion design, production, and marketing, from the smallest start up designers who themselves cut and sewed their first small production runs before they could contract out for production to the largest manufacturers.

To preserve this mix, New York implemented zoning that created a preservation area for manufacturing. The city and building owners envisioned it might transition to a combination of design, sales, marketing, production, and other office uses and, in an effort to achieve the desired balanced mix of uses, the SGCD provided that owners could convert space to higher-rent uses only if they dedicated an equal amount of space for production. Unfortunately, the city failed to provide strong enforcement mechanisms and over time millions of square feet of space were illegally converted.

Cities are again taking up the challenge of preserving a mix of uses. The Greenpoint Williamsburg area of Brooklyn has emerged as a center for the creative economy and artisan entrepreneurship. But the City had already rezoned long stretches of the waterfront from industrial to residential and is now coming to realize that more industrial space is needed to seize these opportunities for business and job creation. The construction of new industrial space is generally not financially feasible without subsidies given New York's high land costs, so the city is offering developers a bonus of additional office space if they commit at least 17 percent of a building to production space for small-scale manufacturers. While the office bonus creates an internal cross subsidy to close the financial gap, the strategy repeats the Garment Center challenge of how to enforce use restrictions when there is a tremendous differential in rents between the uses.

Garment district in NYC
(Source: Wikipedia entry)

Enforcement and Compliance

The city's proposal does not include enforcement mechanisms for ensuring that these spaces are used for production and both the residential community and industry advocates are proposing a variety of additional tools to improve enforcement. One proposal is to require an "administrative agent" to assess compliance annually, similar to the assessments which are done in the affordable housing field to ensure that developers are following income guidolines While this might assure minimal compliance, e.g., proving that a tenant is "industrial," it does not necessarily maximize the public policy objectives of creating a density of high quality jobs, fostering synergy between the tenants and other community assets, or enabling resident employment.

Another proposal is to require that a nonprofit organization rent the production space and then sublease it to Makers and manufacturers. The tenanting or curatorial decisions would be guided by the policy objectives.

San Francisco is implementing another model, which encourages that the production space be owned by a nonprofit organization that will curate the tenants. As in New York, San Francisco has an extraordinarily tight real estate market and the construction of new production space is not financially feasible. They, too, are using a cross-subsidy model and recently enacted legislation which allows office development in an industrial area only if 33 percent of the building is used for production. During the special permit application for Hundred Hooper, the first mixed office/production building to be developed, the Planning Commission negotiated that 56.4 thousand square feet of the total 199.2 thousand square feet of production space be owned by PlaceMade, a nonprofit industrial real estate developer. The use of a nonprofit both maximizes the potential to achieve the policy objectives of providing affordable space for small scale production, it also creates a yardstick for assessing compliance in the 142.8 thousand square feet of space that remains in the developer's control.

Nonprofit Developers

Strategies that use a nonprofit developer to preserve space for production may seem more appropriate for cities where real estate pressures create a tremendous incentive to violate zoning. But they offer another set of advantages in weak market cities: they often create a vehicle for the use of a city's capital funds to subsidize a project that "primes the pump" for development in an area. They also generally have strong ties to a community and can strengthen the pipeline that provides training and job placement for neighboring residents. Finally, real estate development is a long-term strategy and as cities across the country are witnessing, what seemed like a weak market only a few years ago is now robust and actually threatening the diversity that made it attractive in the first place. Better for a city to anticipate change and put in place mechanisms that preserve diversity moving forward than having to play "catch up" and seek to provide space in a strong real estate market later.

HACKING POLICY TO MAKE REAL ESTATE MORE AFFORDABLE

BY KIM-MAI CUTLER

Reporter, TechCrunch

Given the immense amount of market pressure to convert former industrial warehouses into offices and condominiums, cities that want to preserve some remnant manufacturing facilities as Makerspaces need to act deliberately and quickly.

After American manufacturing reached its peak share of employment in the 1970s, factories both decentralized their workforce out to the suburbs and began moving jobs overseas. This left industrial warehouses in disuse in major American cities like New York. Famously, artists began illegally converting some of these mid-size and smaller factory spaces into lofts in 1960s SoHo. As quickly as a decade later, these lofts became highly sought after as a form of upscale living. Artists, who had once sought to legalize loft living in the 1960s, made a political U-turn and tried to prevent legitimizing them to broader swaths of real estate investment capital to hold property values down.

Likewise in the commercial arena, warehouses achieved an "industrial chic" aesthetic that became coveted by younger companies and startups in the first dot-com boom in San Francisco.

Facing the loss of these spaces by the mid-2000s, San Francisco invented a new zoning category called PDR, or Production, Distribution, and Repair.

"It was a branding move," explained Steve Wertheim, who works in the city's planning department. "We needed to protect industrial use but we needed a term that covered a broader range of activities including the arts."

The word "industrial" didn't adequately describe the new non-housing and non-office uses the city wanted to protect and it evoked too many of the heavy uses from the early twentieth century.

"Back then, we had these industrial districts from the 1950s when industry was smelting lead and grinding animals—just nasty stuff—and it was a huge part of our economy," Wertheim said. "Now industry is so much more benign."

So PDR, what Wertheim called a "plannery-jargony" term, became the umbrella definition. Over time, as office spaces have shot up to per-square-foot costs that exceed those of the late 1990s and early 2000s, this category of space has come to trade at a 30 to 40 percent discount.

In certain neighborhoods, the city also created incentives to produce new production, distribution and repair spaces on large vacant sites and re-developable land. Around parts of San Francisco's Mission District, real estate developers who build large enough office buildings must set aside one-third of their newly created space for PDR.

Inadvertently, this has enabled more capital-intensive hardware startups, which have experienced an unexpected resurgence in the early 2010s thanks to crowdfunding platforms like Kickstarter, to find space in competitive San Francisco. Nomiku, which makes a home sous vide kit, was able to find affordable space in San Francisco and then partner with 3D printing houses in Oakland to rapidly prototype new versions of their device instead of building and manufacturing in Shenzhen, China, where they had been based before.

It has also enabled larger manufacturers like Heath Ceramics to co-exist alongside upscale coffee shops and startup offices.

However, at times it can become a contentious issue. In 2014, Pinterest, the software platform that is popular among women for collecting images of fashion and home design, tried to move into production, distribution and repair space that was already occupied by furniture retailers. After an uproar, city supervisors quickly nixed the deal and Pinterest had to look for space elsewhere.

Wertheim says that the city upholds the category through complaint-based enforcement, which means they can sometimes miss or overlook uses that don't necessarily live up to the category.

Preserving Independent Retail
San Francisco also has a long history of resisting standardized chain stores and supporting independent retail through intentional quirks in its zoning code. In 2006, voters passed a "formula retail" proposition that requires stores with more than 11 global locations to go through an extra layer of review in certain neighborhoods. While this has prevented bigger chains from eating up retail space that could otherwise go to local businesses, it hasn't prevented the boutique-ification of neighborhoods like Hayes Valley, which now have independent, but very upscale, small shops.

The city PDR zoning also allows businesses like Heath Ceramics to sell their product in parts of their factory, which lets consumers buy directly from manufacturers in their Makerspaces.

Guarding against Inadvertent Loopholes
Sometimes, land-use policies intended to protect artists and Makers can go awry. In the late 1980s, San Francisco's planning commission enacted new rules to prevent the loss of artists and art-related businesses by creating a zoning category that allowed housing in industrial areas, but only for people who worked and lived in the same space.

It was designed for artists, but in the mid-to-late 1990s tech boom, real estate developers began using it as a way to build thousands of condos in SOMA, the Mission District and Potrero Hill in a loft style with 17-foot ceilings, large windows, and roof decks. By 1999, the Planning Commission began restricting construction of these live-work spaces, seeing that they weren't being used by their originally intentioned target audience but rather by young professionals and office workers.

Inclusionary Zoning for Affordability

As remaining parcels of land in highly demanded cities get rapidly converted to office and residential use, there is now an extraordinary amount of pressure that is working against preserving socioeconomic diversity in neighborhoods. At the same time, both federal and state funding for subsidized, affordable housing has disappeared across the country. In California, state funding for affordable housing has declined by more than 60 percent since the re-development agencies were dissolved in the wake of the last financial crisis.

One of the last remaining tools in the bucket is inclusionary zoning, which uses market-rate development to fund a small number of subsidized units for lower-income workers. Since 2012, San Francisco has required developers to either build 12 percent affordable housing on-site, or fee out of the program by paying 20 percent of a project's cost into a citywide fund. Voters recently raised that requirement to 25 percent. New York City Mayor Bill De Blasio also recently adopted a mandatory inclusionary housing policy that relies heavily on the concept to either build or rehabilitate 200,000 units.

Inclusionary zoning is not a panacea; it is one tool out of many that need to be used. In San Francisco, this policy produced fewer than 2,000 units between 1993 and 2014, but it is one of the last remaining sources of funding for building income-protected housing given the loss of state funds.

The units these programs build, called "below market rate" or BMR, units are income-restricted to people earning a certain percentage of area median income. Typically, these are for low-income residents, although the housing affordability crisis has become

DESIGNING SPACE FOR INNOVATION, CREATIVITY, AND ENGAGEMENT

subsidizing the production of middle-income units. These units are awarded by very competitive lotteries and they cannot be resold for a higher profit.

BY HEATHER KING

Managing Director, Eight Inc.

In recent years, interest in innovation has proliferated beyond universities, NASA, and corporate R&D. Cities are getting into the action, looking to support this next generation renewal through citizen engagement and private/public partnerships.

Eight Inc.—as a strategic design firm—is increasingly active in

Source: Eight Inc.

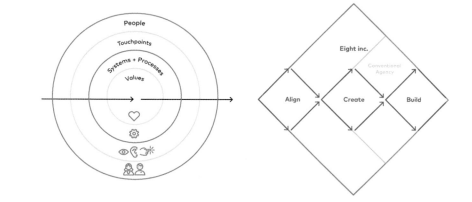

innovation that happens where the city is the unit of change.

Future Cities Catapult Innovation Centre, UK as a Public/Private Hub

Most recently, we worked with Future Cities Catapult (FCC), to create a global center of city innovation. FCC wanted to develop the Innovation Centre, a space for business, universities, and government to develop urban solutions.

FCC Innovation Centre is the antithesis of the traditional office. It was conceived as a platform for experts from various backgrounds and disciplines to work together on a variety of projects. Some collaborations involve small, focused teams. Other efforts are more complex and require a larger team and more continuous space.

The Eight Inc. team used an outside in, inside out approach to develop a human-centered design plan for Future Cities Catapult.

The strategy and implementation effort spanned three distinct phases: align, create, build:

1. Align around the experience needed by an existing and emerging audience.

2. Create the product through a highly collaborative, agile, democratic prototyping process.

3. Build and execute using best-in-class delivery.

FCC required thoughtful orchestration of physical spaces to support an array of team sizes and collaborations. It also invokes novel uses of technology to support problem solving and to showcase invention. The exhibition/demonstration area consists of analog displays and a digital interactive table with an adjacent video wall. The interactive table holds a large touch-screen. Physical objects drive the interaction: as the objects are moved to different areas of the table they trigger content both on the table and on the video display.

The successful outcome of Catapult is highly rooted in this collaborative three-phase process: align, create, build. At Catapult, FCC

executives, partners, and stakeholders were engaged; Eight Inc. worked together with stakeholders to conceive its product and present it to an executive team. This was an inclusive and iterative process honoring inputs from the myriad of stakeholders. This singular focus on human needs and the radical inclusion of users and partners is central to driving a successful Catapult Centre experience. Attention and support from the top levels on through the service department and most needy customer is equally critical.

Small Town Innovation in Minden, Nevada

While many innovation districts and incubators are springing up in large- and mid-sized cities across the world, we are also seeing Making activity combined with retail springing up in small, rural locations as residents, small businesses, and landowners work to proactively direct the future of their communities. This is urban renewal and civic innovation writ small, yet it is popping up in towns and rural communities across the U.S. and beyond. In April 2016, the Berkshire area in rural Massachusetts and Connecticut announced the launch of an "innovation lab" to cultivate regional innovation renewal.

For the past few years, Eight Inc. has been working with rural landowner Chris Bentley in Minden, Nevada, to create what is essentially an urban renewal project. Minden is a frontier town at the Eastern edge of the Sierras, just over the hill from the tourist mecca of Lake Tahoe basin. Bentley is leading an effort in the community to historically and sustainably "reinvent" a cluster of structures at the center of town. Rather than house unglamorous and somewhat toxic gambling outposts, they are working to develop a core commerce that has financial viability and makes best use of land, crops, and location. The mill and creamery structures will become a boutique whiskey distillery, with an adjacent visitor center that provides learning and participation for both locals and visitors. The interior and exterior spaces are integrated; the façade facing Main Street invites. It is at once an important economic engine for the region, and a place to gather and learn.

The Minden project will impact the community profoundly. Locals will have a new a space that invites them to recreate, appreciate their surroundings, and encourage mingling with visitors. This, in turn, will promote the interconnectedness of their remote location with the larger world community. They will reap the economic benefits of more jobs and more locally-based business tax revenue. With the preservation of historic structures, townspeople can celebrate their heritage. The strict adherence to sustainable agriculture and business practices will yield healthful benefits and set the bar for other redevelopment efforts. This innovation district is the embodiment of urban sustainability in which the diverse factors that will sustain the town into the future–the economic, social and environmental benefits–are considered and addressed. It is an urban biodiversity that is characteristic of the current renewal and at the heart of sustainable cities. In essence, it is similar to the well-known and successful High Line Park in New York City and the Ferry Building in San Francisco. Reviving and reinventing important public-private spaces drives a long term, sustainable benefit for the community.

Conclusion

The acid test for a successful outcome is whether people get excited, really excited, about a product or experience and make it their own. We believe that Maker Cities can effectively do this. Maker Cities can create a community hub around a Makerspace, an innovation center, or an urban renewal project and can tap into a deeper emotion. To encourage community engagement, the trick is in the design and in the execution. It is easier to envision a place that feels innovative than to make it happen. Any initiative requires the vision and support of great leaders, a laser focus on all elements of the human experience, and best in class implementation.

IMPLICATIONS FOR CITIES

Innovation Districts could be the answer for your city when looking at how to make room for Makers to work. The best such districts enable Makers to knock shoulders with other creative people (designers, architects) as well as with those in the tech community. The challenge is creating a diversity of spaces at differing rent levels and not allowing high rent uses to squeeze out lower rent uses that both stimulate innovation and broad job growth. The great irony is that a vibrant creative business sector requires stable real estate. Pratt Center argues that zoning can encourage a mix of space and vesting management, and ownership in a mission-oriented nonprofit can add stability as well as promote tenant curation that maximizes job growth, advances equity and inclusion, and facilitates resident employment.

Innovation centers and labs are popping up everywhere. But building one isn't as easy as it sounds. You need a champion who is laser-focused on building the user experience around the people who will work in your innovation center. Each touch point needs to be carefully designed to have a visceral impact on the senses. Space needs to be carefully programmed and planned for casual and accidental interactions.

Be a good client when commissioning creative space to be built from scratch. Supportive. Not beholden to preconceptions. Willing to throw out "industry standards" and try something new and different. Committed to "getting it right" even if "it" is ineffable.

Working with real estate is like everything else in the Maker movement. Commit to rapid iteration, much experimentation, and don't bother arguing about the conceptual. Build a prototype instead. Bring in experts early to point out challenges and develop solutions. Get the team that is working on a real estate development together weekly if you can, as you'll find that a week then gives you the same productivity other teams get in a month. Clients need to commit to reviewing materials and providing a decisive reaction quickly so as to keep the project on track and the team from getting buried under an avalanche of delayed decisions.

Policy hacks look promising, but we have yet to crack the code on how to make space inside our gateway cities affordable for Makers to live and work. We need to spin up more policy hacks that experiment with how to make both work and living space more available to Makers at prices they can afford. Perhaps a civic hackathon could help here; see Chapter 8 on Civic Engagement for details.

RESOURCE

QUICK
WIN

IMPLEMENTATION
ADVICE

POLICY

BIG
IDEA

MAKER METROPOLIS

BY JENN SANDER

The most Maker City of any city is without doubt Burning Man, a remote place where nothing exists until its citizens create it—once a year, and for only a week.

PHOTO CREDIT
Scott London

Every August, Burning Man rises in Nevada's barren Black Rock Desert, a thousand square miles of wilderness with no water, no electricity, no services, nothing really but a vast playa (a dried antediluvian lake bed) in one of the most beautiful and harsh environments in America.

It was this stunning blank canvas that initially inspired people to come together to dream up a city where they could follow any whim they wanted to try. A do-it-yourself community fueled by the love, triumph, fun, and failure of creating the world they wanted to live in. It has become a Maker metropolis that real cities are learning from today. Ultimately Burning Man is about place making with the entire city built "by the people, for the people." This sense of freedom and engagement leads to civic agency that has practical implications for how cities can embrace the spirit of making—and become a maker city.

The community started out organically in 1986 and has evolved into a city imagined and reinvented by its 70,000 participants each year. It's a metropolis of creatives, Makers, performers, explorers, and inventors with its own post office, airport, radio stations, newspapers, volunteer rangers, a DMV (Department of Mutant Vehicles), and a plethora of large-scale interactive art—with no branded content at all. It is a gift economy where nothing is bought or sold, so participants must be self-reliant and at the same time come together to learn, create, and thrive. When the event is over, the whole thing vanishes without a trace. Black Rock City, the largest annual Burning Man gathering, is meticulously cleaned up and packed out by all of its participants at the end of the week.

Key Principles of Engagement:

In 2004, Larry Harvey, one of the organization's Founders, assembled Burning Man's 10 Principles as a reflection of the community's ethos and culture to help nurture organic offshoots around the world. In no way prescriptive, these patterns of behaviors serve as guidelines for affiliation and help encourage emergent activity for thousands of participants, theme camps, and art project teams who embrace it as a framework.

OUR ETHOS IS BUILT ON THE 10 PRINCIPLES

A SET OF COMMONLY UNDERSTOOD VALUES REFLECTED IN THE BURNING MAN EXPERIENCE

Radical Inclusion: Anyone may be a part of Buring Man. We welcome and respect the stranger. No prerequisites exist for participation in our community.

Gifting: Burning Man is devoted to acts of gift-giving. The value of a gift is unconditional. Gifting does not contemplate a return or an exchange for something of equal value.

Decommodification: In order to preserve the spirit of gifting, our community seeks to create social environments that are unmediated by commercial sponsorships, transactions, or advertising. We stand ready to protect our culture from such exploitation. We resist the substitution of consumption for participatory experience.

Radical Self-reliance: Burning Man encourages the individual to discover, exercise, and rely on his or her inner resources.

Radical Self-expression: Radical self-expression arises from the unique gifts of the individual. No one other than the individual or a collaborating group can determine its content. It is offered as a gift to others. In this spirit, the giver should respect the rights and liberties of the recipient.

Communal Effort: Our community values creative cooperation and collaboration. We strive to produce, promote, and protect social networks, public spaces, works of art, and methods of communication that support such interaction.

Civic Responsibility: We value civil society. Community members who organize events should assume responsibility for public welfare and endeavor to communicate civic responsibilities to participants. They must also assume responsibility for conducting events in accordance with local, state, and federal laws.

Leaving No Trace: Our community respects the environment. We are committed to leaving no physical trace of our activities wherever we gather. We clean up after ourselves and endeavor, whenever possible, to leave such places in a better state than when we found them.

Participation: Our community is committed to a radically participatory ethic. We believe that transformative change, whether in the individual or in society, can occur only through the medium of deeply personal participation. We achieve being through doing. Everyone is invited to work. Everyone is invited to play. We make the world real through actions that open the heart.

Immediacy: Immediate experience is, in many ways, the most important touchstone of value in our culture. We seek to overcome barriers that stand between us and a recognition of our inner selves, the reality of those around us, participation in society, and contact with a natural world exceeding human powers. No idea can substitute for this experience.

In many ways, it's these principles that provide the permission for social experiments to take place, and for society to be prototyped and re-designed. These principles are often in tension with one another: Radical self-reliance and Radical self expression are all about the individual. Civic Responsibility and Radical Inclusion are all about the community. It's this constant interplay between collective governance, individual novelty, and emergent activity that is the scaffolding that makes Burning Man such a creative Maker City.

"Out of nothing, we created everything," Burning Man Founder, Larry Harvey.

Burning Man and the Maker Movement:

The event has received much attention for being a petri dish for new technology and art fueling the Maker Movement. However, Burning Man culture is less about the artifacts and more about the mentality of being a Maker and the values of participation, experimentation, and play, with a point of view that is focused on the art first. Beauty and curiosity are put on the same level as functionality and creations are often burned at the end of the week.

A key principle highlighting the Maker spirit of the city is participation. There are no "observers" at Burning Man. The belief is that to be there and receive gifts everyone must participate. Giving permission to everyone to create makes for a heightened sense of civic responsibility and communal effort. These are ideas that enhance any city.

How the City Works:

Preparations for Burning Man take months and involve scores of volunteers, who work together to create the different camps or neighborhoods where people will live as well as the large-scale and interactive public artwork that will decorate the desert floor.

The principle of Civic Participation is furthered by the principle of "leave no trace" - nothing on the desert floor when you arrived; leave it that way. As in real cities, Burning Man uses data and maps to monitor all this. Volunteers develop a "Matter Out Of Place (MOOP) Map" which helps to grade camps on how well they remove traces of anything. This is an example of data about the community being used to drive appropriate behavior.

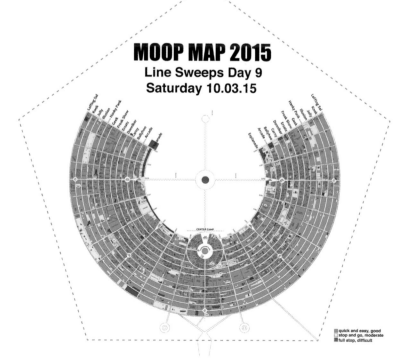

No cars are allowed inside Black Rock City.

Instead, roving works of art double as public transport. These "Mutant Vehicles" take people around the inner ring of Black Rock City as well as into the City Center. Every year, Mutant Vehicles get more elaborate and ambitious and require ever greater feats of engineering and imagination.

Black Rock City is Designed in a Radial Fashion

El Pulpo Mecanico was built in Arcata California, Humboldt County. Designed and built by Duane Flatmo along with his friend Jerry Kunkel who masterminded the electrical systems and flame effects.

At the center of the radius is "the Man," an iconic sculpture which can tower 100 feet and is burned in ceremonial fashion on the last night of the week-long festival. Streets are mapped out only from 10:00 to 2:00, leaving open space for views of the mountains and to create lines of sight to the "Playa," where most of the large-scale art is located.

At 6:00 on the map, you'll find downtown, the heart of the city known as Center Camp, a 45,000-square-foot shaded structure informed by one of the most important influential models of civic design, the Italian Piazza. This comfortable gathering place encourages chance meetings, deep conversations, and new ideas - much like the historical cafe culture that has given birth to everything from existentialism to beat poetry. It is extremely important to have this feature of urban life, this "third space," as a special meeting and social hub, at Black Rock City.

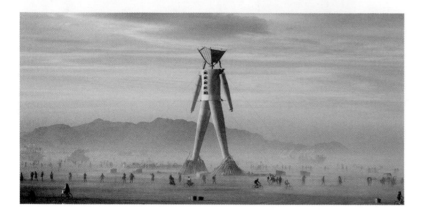

"The city's design helps maintain the sense of expansive, infinite possibility that is natural to the landscape, and also encourages people to voyage out and discover new things. Everyone is attracted to the center and makes their pilgrimage to the Man," says Burning Man Co-Founder Harley Dubois, making random collisions inevitable and part of the flow. Serendipity is literally part of the "way" here.

The Growth of the Movement:
Burning Man culture has expanded worldwide with 270 volunteer regional representatives who mentor the community in 35 countries and 130 cities and towns. These communities, embedded in American and global communities, are a significant source of inspiration, volunteer capacity, and Maker know-how.

People take the immediacy and civic responsibility home with them. Building Burning Man gives people a sense of their own agency and makes it clear how radical self-expression and civic engagement ultimately leads to a better life. It's this civic pride—a more engaged form of citizenship not just in Black Rock City but in the real world—that underlies the global voluntary association and network of social trust known as the "Burner" community.

Emergent Activity in the World

David Best, for example, is famous for creating architecturally elaborate temporary temples across the globe. He never imagined that what began as an experiment in the desert would become a worldwide phenomenon. As a result of his initial structure at Burning Man in 2000, the Temple is now a permanent installation built by different guilds, volunteers, and collaborators who submit proposals of their interpretation for what a Temple might be. Being selected to build the temple is a high honour; it has helped many new talents establish themselves internationally.

In 2015, the Artichoke Foundation commissioned David Best to work with the community of Derry, Northern Ireland to build a Temple on a remote hilltop overlooking the city. The project was designed to unite the community, celebrate togetherness, and help citizens and residents of Derry heal from its traumatic history of conflict. A crew of temple builders came from California to work with local students and adults - with the support of local fablabs - as a form of skills training and development.

Burning Man's Civic Arm

Burners have also put their Maker skills to use healing communities through an organization called "Burners without Borders" (BWB). Begun as a response to Hurricane Katrina, volunteers with

Burners without Borders provided over $1 million worth of reconstruction and debris removal over an eight-month period.

Burners without Borders spawns new projects each year. One shining example is Communitere, a dynamic and sustainable organization that provides Makerspaces to communities that have suffered from natural disasters. This allows people to create the relief efforts they need for themselves, tapping into their own skills and resiliency and applying their knowledge of what they and their neighbors actually need right now. Communitere now operates in Haiti, Nepal, and the Philippines, providing space to NGOs like Field Ready, who is pioneering the use of 3D printers in disaster areas.

Lessons for the Maker City

Burning Man has been called a "permission engine" that allows for both radical individual expression and rich community collaboration. It's a construct built on trust and shared values that allows its constituents to create all the facets of the city they want to live in.

The 2016 theme for Burning Man was "Da Vinci's Workshop," inspired by the Italian Renaissance when an historic convergence of inspired artistry, technical innovation, and enlightened patronage launched Europe out of medievalism and into modernity. The era was known as "The Age of Discovery," when the ideas of da Vinci, Michaelangelo. Copernicus, and Columbus reimagined education, production, science, and politics. Like today it was an era of fundamental reframing. Renaissance workshops were forerunners of today's Maker and innovation spaces—a breeding ground for new ideas that also helped them become reality. Ateliers were established in the Renaissance and saw participatory and vocational knowledge as the core of value creation. In the safe environment of the atelier, established artisans could spot and mentor new budding talent and ideas—networking them together across disciplines, thus fostering new potentials. The ateliers' major themes resonate with the innovation economy today: turning ideas into action, cross pollinating art and science, and unleashing human imagination.

The Burning Man Project

Rooted in the values expressed by the Ten Principles, this culture is manifested around the globe through art, communal effort,

and innumerable individual acts of self-expression. As a newly formed nonprofit organization, the Burning Man Project is focusing on offering artists, changemakers, cities, nonprofits, and civic activations a path to becoming agents of change to manifest new ideas in the world.

Participating in Black Rock City or other regional Burning Man events is just one way to get involved with the community. Wherever you are and whatever your skills, there are plenty of exciting opportunities to volunteer, collaborate, submit ideas, and get involved with the global efforts of the Burning Man Network via Burningman.org.

→ **Jenn Sander** is the Global Initiatives Advisor, Burning Man Project

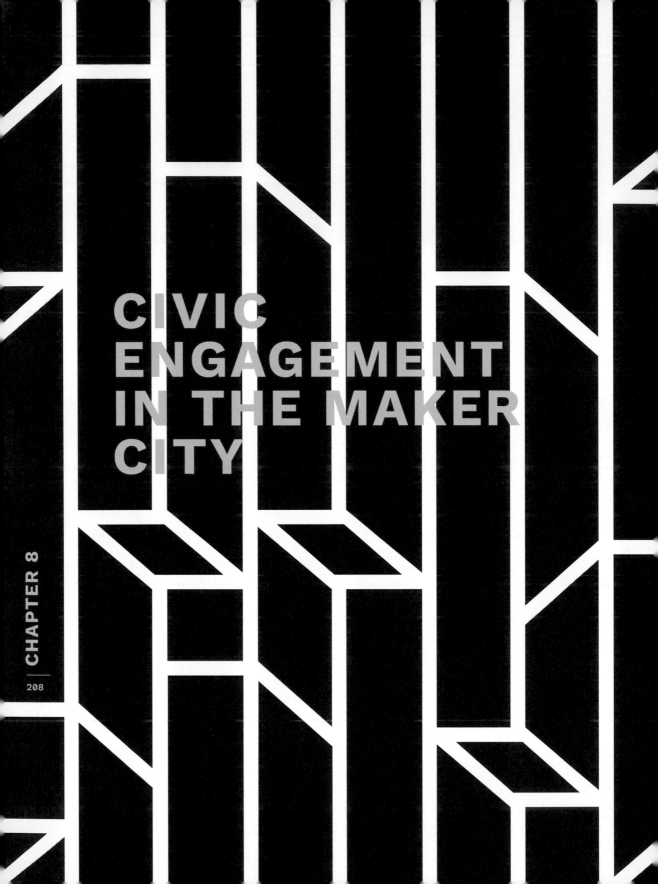

CIVIC ENGAGEMENT IN THE MAKER CITY

WHY A MAKER CITY IS OPEN TO ENGAGEMENT

The greatest expression of a Maker City is a citizenry with the skills, ideas, and will to make the city itself.

"Over the next decade and beyond, cities will continue to capture the imaginations of people around the globe—drawing in nearly a billion new urban dwellers and spurring the largest and fastest reinvention of our built environment in human history. As cities grow, traditional top-down approaches will be insufficient to meet the basic needs of citizens or fulfill their aspirations. Already, citizens and Makers are stepping in to fill this gap, leveraging emerging technologies to remake the stuff, services, and systems of urban life itself."—**Ben Hamamoto, Institute for the Future (2014)**[1]

The United States has a long tradition of communities coming together to make things collaboratively, a tradition that started with barn raising in our rural communities and was reinterpreted more recently with Habitat for Humanity bringing together volunteers from all walks of life to build affordable housing for those who can least afford it. The Maker examples cited in this chapter follow in that tradition.

In this chapter, we'll talk about how Makers are continuing, and building on, a long tradition of civic engagement, to take what is working, build on it, and in the process work, to correct parts of our cities that are not functioning as well as we would like.

Examples will be drawn from tactical urbanism in three different areas:

→ Housing advocacy (Seattle)

→ Open-space design (San Francisco)

→ Urban prototyping (San Francisco)

We'll also bring in examples from New York City, Boston, and Chattanooga, as we talk about how cities are closing the digital divide and engaging young people in civic action.

One of our underlying themes here is that good ideas flow from experiments. Some experiments work, some don't, but learning from failure and then improving on it is an essential part of innovation.

Another theme is that technology—its culture, methods, and tools—can work both as a kind of glue and as an accelerant when it comes to change.

HOUSING ADVOCACY IN SEATTLE: THE IMPORTANCE OF STARTING SMALL

In the Maker City it is not enough to make something that is innovative. To make innovation stick, Makers must learn how to work across sectors, with established city leaders and nonprofit organizations to build acceptance for what they've made. One way to do so is to tie Makers' work to the bigger agenda at work inside their city.

A group of Makers in Seattle did this to good effect. Mayor Murray declared a "state of emergency" around homelessness in late 2015. Makers there were able to create a village of 14 tiny homes that was well received by city officials.

The Seattle project is called "Tiny House Village" and is championed by the Low Income Housing Institute, a nonprofit that both develops and builds housing for low-income people, working closely with city officials to coordinate supportive services and obtain land use rights and permits. So far there are three tiny house villages in Seattle.

Tiny House Village,
Seattle WA

This is an example of the Maker mindset applied to civic engagement at its very best.

The 14 tiny homes that were built represent an experiment in a new form of housing for the homeless that can be built very quickly and at minimal cost. The advocacy group that built these homes did not ask for formal permission and did not seek permits. Yet the model was and is highly replicable. Tiny home villages can now be found in Fresno, California; Eugene, Oregon; Madison, Wisconsin; and Austin, Texas. (Source: **Christian Science Monitor, 2016**[2])

Tiny houses are an example of what urban planners call "Tactical Urbanism:" a project purposefully built to solve a problem and fielded as an experiment versus created as a permanent fixture.

Tiny homes can help relieve the problem of homelessness temporarily by giving the homeless a place to live that offers them both dignity and privacy.

Recently housing advocates in Oakland launched Pop-Up home build days, which created Tiny Homes on Wheels for the homeless. Of course, tiny houses are not a panacea. Experts in NYC think of tiny houses as more of a "niche solution" to housing the homeless.

Our point is this: Makers can create solutions to problems as a set of experiments very rapidly. Some experiments succeed, others fail, but Makers' willingness to experiment provides a positive example, one that encourages city officials and established nonprofits to work together with the community to come up with more permanent and broad-reaching solutions.

RESHAPING PUBLIC PARKS: MAKERS KEEP IT REAL

Another example of tactical urbanism comes from San Francisco, which pioneered the concept of turning parking spots into small parks called "parklets." Parklets convert private auto parking space into additional public gathering space.

Exploratorium Parklet

The original Parklet concept was created by Rebar Art and Design Studio (now Morelab and Gehl Studio) in 2005 by transforming a single metered parking space into a public park in San Francisco. The Parklet launched as an open source project in 2006, when Rebar had the first ever Park(ing) Day and it has since become an international event. Park(ing) Day occurs annually on the third Friday in September, in hundreds of cities around the globe.

Open sourcing the Parklet idea allowed the concept to scale rapidly and have broader impact which inspired Mayor Gavin Newsom to support an effort to make an official Pavement to Parks program in 2009. This in turn allowed parklets to become an official part of San Francisco's landscape. The parklet program has continued under Mayor Ed Lee, in an effort to reduce automobile traffic, turn parking spots into public plazas, and convert excess roadway into places for people to congregate.

The idea is simple. The city provides the space and the review process; private entities (merchants, community organizations, individuals) submit a site plan, step in and implement the parklet, and are responsible for maintenance, insurance, and opening up the parklet to the public. About 42 proposals were received in 2009 from private entities, which resulted in 38 parklets being built in 2010.

One of the parklets on Valencia Street between 23rd and 24th Streets, in the heart of San Francisco's Mission District, was built by young people from the area working in conjunction with a grant from the National Science Foundation and the Exploratorium. The Mission District of San Francisco has historically been a poor, working class, Spanish-speaking neighborhood. More recently, the Mission has been in the news because rapid gentrification is displacing families that have lived there for generations.

San Francisco Boys & Girls Clubs (BGC) and a team from the Exploratorium's Studio for Public Spaces took charge of the parklet, taking over two years to build it. It was structured, quite literally, as a science experiment; one of the goals of the parklet, called Casa Ciencia Pública: Agua, is to encourage discussions around water and sustainable water use.

Parket in the Pacific Heights district of San Francisco

The people who worked to build the Casa Ciencia Pública: Agua parklet on Valencia Street did not necessarily call themselves "Makers." Only in retrospect can we think of the woodworkers, welders, ironworkers, fabricators, and construction workers who stepped up from the community to help build this parklet as Makers.

The parklet concept was immediately popular. Most were successful, but some did not go so well, striking the wrong chord with city officials or members of the community and were ultimately taken down. One city supervisor found the parklet below, designed and built by a design firm with a national reputation at a cost of $40K, objectionable for stylistic reasons.

(This parklet remains in place ... despite some initial resistance.)

Other parklets contained structures that people in the neighborhood found too commercial, interfered with traffic in unexpected ways, or proved to be too "cookie cutter."

As of 2014, there were 43 parklets in San Francisco, many beloved features of their neighborhoods. (Source: **Curbed**[3], 2014)

The Maker movement and tactical urbanism share a passion for experimentation, a willingness to plan for failure, and a reliance on tinkering and rapid iteration to achieve success.

In 2011/12, city officials expanded the parklet website to accommodate the high level of interest in the program and to help other municipalities around the country who had shown interest in developing their own parklet programs. The **website** provides the following:

pavementtoparks.org/parklets

→ **A detailed roadmap as an infographic** so that everyone involved in the parklet program can see and understand time frames and next steps at every point of the process

→ **A materials list** making it easy for anyone with DIY skills to participate in building the parklet

→ **A policy framework** that explains why the city is interested in seeing parklets flourish and so that merchants and Makers can create proposals that are responsive to the city's needs

→ **Detailed permitting requirements** specifying exactly how proposals from the merchants will be viewed

Makers are ideal for fabricating parklets. They work inexpensively, make innovative use of materials, and, to the extent they come from the neighborhood, know what is appropriate to build there. While there is no "average cost" for a parklet, anecdotal evidence suggests that costs can be reduced through the involvement of Makers.

URBAN PROTOTYPING—HACKING OUR WAY TO BETTER STREETS

Conventional urban planning processes make it difficult to implement great ideas for civic improvement. **It now takes four times as long to move along civic infrastructure projects than it did in the 1970s**[4]. This might well be a reaction to planning czars like New York's Robert Moses "titan of the skyline" who was famous for his legendary speed and efficiency (and the ability to route around any agency in his way), but also for running rough-shod through neighborhoods with roads and slum clearing projects without much interest in community input or participation.

"The traditional model of city-making has historically involved experts with a definitive, long-term plan executed over time. The issue with that is that culture changes faster than infrastructure; we've surpassed our ability to keep up. One of the consequences is that we're left living in cities we planned 50 to 60 years ago." (Source: **Interview with Blaine Merker, Gehl Studio on Gizmodo, 2013**[5])

Could a model popular in the tech industry–experimentation and rapid iteration–be applied to the process of city planning?

Urban prototyping complements city processes by rapidly testing, sharing, and scaling new citizen-sourced projects that improve civic life. The concept was initially developed by nonprofit organization Gray Area Foundation for the Arts (Gray Area) based on several years of producing civic hackathons aimed at the creation of digital civic applications in partnership with city government. You can think of urban prototyping as a multi-year hackathon for the betterment of the public realm.

In 2012, Gray Area partnered with Intersection for the Arts to produce the first Urban Prototyping Festival which took shape around the 5th and Market and 5th and Mission intersections in San Francisco. A total of $25K was invested in 25 prototypes that were developed over three to six months and deployed through a weekend street closure permit. The street festival allowed for 5,000 citizens to experience redesigned sidewalks with everything from new concepts for street lighting to public urinals developed by Makers. The goal of urban prototyping was not only to showcase projects but to also uncover urban planning conventions and permitting processes that were ripe for innovation.

A new permit type was developed shortly thereafter in 2013 called **Living Innovation Zones**, allowing for temporary installations to activate sites along Market Street.

sfliz.com

Market Street is one of the most traveled streets in the city, filled with pedestrians, public transit, and tourists. Many iconic

companies have their headquarters on Market Street including Dolby, Twitter, and Zendesk.

At the same time, walking along Market Street can feel a bit disconnected. There are few opportunities to pause, linger, or converse with others, or to connect with the local community in any way.

Building a Better Market Street is high on Mayor Ed Lee's agenda for San Francisco. Mayors care about quality of life and economic development, two issues that come together on Market Street. In the course of developing a plan for "A Better Market Street," The San Francisco Planning Department and Yerba Buena Center for the Arts (YBCA) with a grant from the Knight Foundation launched the Market Street Prototyping Festival.

The 2015 Market Street Prototyping Festival selected 50 teams to build prototypes of interactive sidewalk installations for placement along the Market Street artery as part of a three-day festival that happened in April 2015. Each prototype had to meet specific criteria in terms of community activation and design, laid out in an open call.

The actual Making by the 50 teams happened over six months, working in partnership with 'District Captains' such as Autodesk and the Exploratorium to build out prototypes. The Market Street Prototyping festival attracted over 250,000 visitors and 25,000 people participated on the Neighborland festival project site, where citizens could comment and vote on the proposed projects.

A detailed evaluation of the impact of the Festival was provided by **Gehl Architects**[6]. The Gehl report showed that the Market Street Prototyping Festival was successful at engaging the community, building capacity in the design community to build more of these types of solutions, and increasing connectivity by encouraging visitors to Market Street to linger and engage with each other. The Gehl Architects' Report is worth downloading and reading in its entirety. All of the projects displayed on Market Street—with detailed instructions on how to build them—can be found on the Instructables site[7].

Today, the festival is positioned to happen annually, bookended with an Incubation program called the Urban Prototyping Research

27 Steps

Market Street has always been the venue to voice, protest for, and celebrate for peace; "27 Steps" reminds us of this journey...

Back to Paper

A paper dispenser of cultural content (poem, short story, drawing, comic strip) created by the community at large. Contribute your art and...

Bench Go Round

Bench Go Round playfully re-imagines public seating to create connection between strangers. Bench Go Round is just what it sounds like, a...

Chime

Chime is an interactive public art installation that uses music and movement to engender a sense of community and wellbeing. Chime is...

Lab, led by Autodesk and Gray Area, allowing for further development to transform prototypes into semi-permanent installations.

All of the Market Street Prototyping Festival efforts will ultimately inform the final multi-million dollar effort to repave Market Street. Through the Planning Department's willingness to crowdsource projects from Makers, and YBCA's efforts to engage the local community, a true experience of being nimble and allowing experimentation is unfolding to create a more vibrant experience for the public.

Urban Prototyping
Projects in 2015

> "By its nature, the metropolis provides what otherwise could be given only by traveling; namely, the strange."
> — Jane Jacobs, The Death and Life of Great American Cities

Urban Prototyping Success Factors

→ **Find one or more local arts groups and city departments to champion your effort**, one that understands the Maker mindset and Maker culture. Choose alliances with city leaders, local organizations and potential funders that can stay on to scale the projects.

→ **Think small.** It's easier to get one small experiment off the ground than to seek permits and permission for something larger and more permanent. Governments seem to be more comfortable with the terms "prototypes" and "pilots" than any term that hints at permanence.

→ **Model your prototyping project as a makeathon or other open competition.** This format is well established and well understood within the Maker community.

→ **Build time into your project for citizen engagement.** One great way to do this is by posting your project on Neighborland for public commentary and voting.

→ **Don't go Guerrilla.** You want to work with city officials, not against them, as much as possible. If you must "go rogue," do so by declaring your effort "just a prototype," to reduce initial resistance.

→ **Plan for obsolescence.** After the completion of Market Street Prototyping Festival, an Incubation program was developed to take the prototypes to semi-permanent installations for a period of up to two years.

→ **Figure out how to measure success.** Ideally, get someone not involved with designing the project to evaluate its success.

→ **Leverage scarcity.** The Bay Lights project is an example of how scarcity can work to your city's advantage with urban prototyping.

Scarcity in Action: The Bay Lights project

This is a light sculpture made up of nearly 12,000 LEDs designed by artist Leo Villareal and affixed to the San Francisco Bay Bridge. The lights shined from dusk until dawn for a two-year period: March 2013 to March 2015.

Engaged citizens lobbied for the project's return and raised money from several sources, including from state coffers (California Arts Commission) and through a crowdfunding site. Today The Bay Lights are back on and there is funding to keep the sculpture running through 2026.

Another lesson learned here is that if you have a big idea, declare it as "just a prototype." There is no way the various agencies would have approved a permanent project on an iconic bridge. But a two-year prototype was approved and went on to win widespread affection from the community.

The Maker City is Made by Everyone

Much of the thinking around the evolving hackathon format came from a pioneering project called Creative Currency, designed to create a baseline of understanding on how to field a civic hack-athon event, one that allows for more civic engagement and a greater amount of time devoted to Making. A traditional hackathon is designed for coders and technologists to show how they can solve a constrained problem in a specific period of time, say 48–72 hours over a weekend. At a hackathon, people work side-by-side with each other, eat a lot of pizza, and at the end of the event showcase what they've built to a set of judges in a competitive setting. A great **how-to guide**[8] to run your own technology-based hackathon is available from Socrata, a company that sells cloud-based solutions for digital government.

Key learnings from the Creative Currency Project were:

→ **Community engagement is essential for any civic hackathon.** Without it, your city will get eager participants with software and technical skills, volunteers who are enthusiastic about urban change, but no one steeped in the community and the changes the community wants to see. The community outreach phase is essential to develop the knowledge and experience to guide civic hackers and link them with community needs.

→ **Ideas can be developed very rapidly in the prototyping phase, allowing participants to suggest and demo new approaches.** Any ideas with merit need at least two months of development and iterative feedback before they are ready to be presented to the community.

PHASE ONE / COMMUNITY OUTREACH

COMMUNITY SURVEY + 30 ON-THE-GROUND INTERVIEWS

COMMUNITY BENEFIT ORGANIZATIONS + LOCAL BUSINESSES + NEIGHBORHOOD LEADERS = HOLISTIC COMMUNITY BRIEF

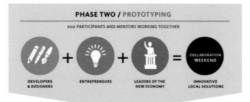

PHASE TWO / PROTOTYPING

200 PARTICIPANTS AND MENTORS WORKING TOGETHER

DEVELOPERS & DESIGNERS + ENTREPRENEURS + LEADERS OF THE NEW ECONOMY = COLLABORATION WEEKEND / INNOVATIVE LOCAL SOLUTIONS

PHASE THREE / DEVELOPMENT & ACCELERATION

2 MONTHS OF ITERATIVE DEVELOPMENT AND FEEDBACK

TECHNOLOGY TESTING + COMMUNITY FEEDBACK INTEGRATION = DEMO DAY / PUBLIC PRESENTATIONS

PHASE FOUR / IMPLEMENTATION & ADOPTION

$15,000 IN SEED FUNDING AND 4 MONTHS OF MENTORSHIP

TOP PROJECTS + FUNDING & SUPPORT = NEW COMPANIES, ORGANIZATIONS, COLLABORATIONS + SHOWCASE AT SOCAP12

Infographic describing how to do a civic hackathon

→ **The importance of demo day, seed money, and ongoing mentoring.** As with code-oriented hackathons, a civic hackathon results in a demo day, where the best ideas are presented to judges and the community, ideally as working prototypes. Seed money and mentoring by a nonprofit organization are ideal to turn winning prototypes into a meaningful test in your city.

OPEN DATA IS AN INVITATION

San Francisco was one of the first cities to open up its data to citizens. Today, about 46 cities and counties across the US have open-data initiatives. San Francisco is a leader here, requiring all its departments to make their data available so as to increase transparency and citizen engagement in local government. To that end, the city of San Francisco makes 524 different machine-readable datasets available through its open data portal: data.sfgov.org. XML has been called the lingua franca of the web for good reason: it is a structured data format that is easy to download, easy to work with, and designed to play nicely with other data sets.

Opening Up the Data in Your City has Unexpected Consequences

It encourages citizen journalism. Citizens analyze the data and blog about their findings on platforms like Medium (long form) and Twitter (micro blogging), gain followers, and encourage action.

It can be used as the basis of public art, to start a dialogue about things happening in your city that need to be changed.

It surfaces problems to city officials in a way that is hard to ignore.

Case Study of Open Data in Action: The Tenderloin Noise and Crime Spotting Projects

Cities are, by their very nature, noisy. The bigger the city, the more likely you are to hear a cacophony of sirens, automobiles, high decibel conversations, and even gun shots.

Recognizing this in 2010, artists at the Gray Area Foundation for the Arts teamed with the engineering firm Arup, the design firm Stamen, and the tech firm Movity (now part of Trulia), as well as coders from the community to deploy noise sensors in the Tenderloin neighborhood of San Francisco.

The Tenderloin is a vibrant but very poor neighborhood, an assemblage of families with children, drug addicts, elderly folks, the disabled, recent immigrants, and people of color all in a 40-block radius. Crime is much higher here than elsewhere in the city. Neighborhood policing here can best be described as a game of whack-a-mole. Police move in and shut down one corner to drug dealers and IV-drug users, only to see that same activity pop up a few blocks away.

The project started with a hackathon, where data sets were made available to all who wanted to participate, including Arup, Stamen, Motivity but also— and most importantly—coders drawn from the community.

The output of the Noise Project was a detailed map that pinpointed areas in the Tenderloin that were hotspots for noise.

At the same time Stamen Design created Crime Spotting to visualize neighborhood crime data. Folks involved in the Tenderloin Noise and Crime Spotting Projects were wary of presenting their findings to folks in the community, as they highlighted the Tenderloin's many problems. Also, the sensors were set up without explicit permission from city officials.

Data from noise sensors deployed in the Tenderloin neighborhood of San Francisco, CA (2010)

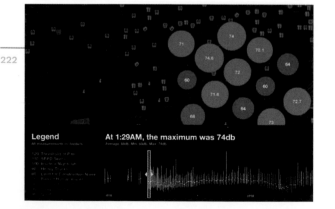

To everyone's relief and surprise, community organizers, police, fire, and other city officials turned out to be excited (very!) by the data uncovered by the project teams working with the noise and crime data. Community activists got involved, lobbied for change, and worked with the police to establish portable police hubs in the 'hood. In short order, neighborhood policing in the Tenderloin shifted from a reactive mode–responding to crime after it happened–to an approach that was more proactive.

Programs similar to the Tenderloin Noise and Crime Spotting Project popped up in Los Angeles and other cities as part of a broader trend toward installing sensors as an input to intelligence-led policing.

This type of reaction to open data, as seen in the TenderNoise and Crime Spotting Projects, has been dubbed the "Read Write" phenomenon by urbanist Adam Greenfield (2010). In the Maker City, data from city agencies gets mashed up with other data sets in ways that bring city problems into sharp focus. Citizen engagement around the data accelerates change. The more transparent you can make your city's data, the better.

Today, there are sensors everywhere in San Francisco. The sensors are low cost and consume very little energy thanks to a partnership with SIGFOX. San Francisco gave SIGFOX the right to deploy its radio network for sensors and devices on public buildings. In exchange the City got the infrastructure it needed to support an entirely new class of applications, based on the internet of things and created by civic hackers. Sensors cost only $1 each and can be used to monitor the city's water pipes for leaks and for predictive maintenance of the city's fleet of vehicles.

Let the Makers Deploy the Sensors
As city spreadsheets get converted to XML by city departments, Makers are continuously finding ways to harness their own real-time data through DIY Sensor Networks. Take the Data Canvas: Sense Your City Project, produced by swissnex San Francisco, Gray Area, and SEEED Studio. One hundred low-cost environmental sensors were deployed by Makers in seven different cities around the world to harness data for research and visualization.

During deployment, each sensor site hosted workshops to teach citizens how to make their own sensors based on an **instructional video**[9] and open source guide produced by Data Canvas. Not only can large companies like SIGFOX or City Governments deploy technology, Makers can use a mesh network approach to scale sensors and other civic infrastructure. **Smart Citizen** is an ongoing project

smartcitizen.me

that allows Makers to use open source technology for political participation in smarter cities.

In short, this is something a Maker City can replicate very easily without too much trouble, providing it thinks ahead to establish an open data policy and gets acommunity-based organizations using that data.

THE IMPORTANCE OF BRIDGING THE DIGITAL DIVIDE

There's a significant digital divide in our cities. Blair Evans of the Brookings Institute commented:

"[In February 2016], the New York Times ran an **article**[10] on how the digital divide particularly affects schoolchildren, creating what they termed a 'homework gap.'"

The article illustrates vividly that today a person's full participation in the economy and civic life requires connectivity.

Maker Cities recognize this and work to get more citizens engaged by providing ready access to low-cost, high-bandwidth internet connectivity.

To do this, New York City started small and worked up from there to build an ambitious program offering. One of the first programs started by connecting just five community rooms inside public housing run by NYCHA (New York City Housing Authority) to the internet. A second program brought a digital van into NYCHA housing developments so that residents could do their homework, look for jobs, and connect with city officials to advocate for change.

Maya Wiley has been involved in programs to close the digital divide in New York since 2014 as an advisor to Mayor deBlasio and head of the task force he created to expand affordable broadband access across all five boroughs. She notes:

"Technology, like broadband access, is literally like water or electricity was at the turn of the century—you can't really do any of the things that build opportunity without it. If we're going to produce more and better-paying jobs, if we're going to improve educational outcomes, if we're going to make sure people can engage with the city effectively … folks have to have the technology and the ability to utilize it." (Source: City Lab, 2015[11])

A list of programs happening in New York City alone to bridge the digital divide and make broadband utilities available to all citizens, just like electricity, water, and heat includes:

NYCHA Digital Van

→   NYC Connected Communities

→   NYCHA Digital Van

→   NYCHA Community Computer Centers

→   Community Technology Centers

Each program started out small and many started out as grass root efforts versus top-down initiatives driven by the city. But the sum of all the programs together adds up to a New York City well on its way to making broadband a municipal utility. Under the DeBlasio administration, New York is deploying its most ambitious connectivity program yet: converting 7,500 former pay phones across the five boroughs into new structures called LinkNYC which will provide free WIFI, phone calls, device charging and access to city services, maps, and directions.

Similar programs are happening in Kansas City, Missouri, through an information partnership with Google/Alphabet Fiber; in Lafayette, Louisiana; and in Chattanooga, Tennessee.

We want to zero in on Chattanooga, Tennessee as an example of what is possible. Chattanooga is home to "The Gig," a broadband offering with internet access as fast as one gigabit per second—about 50 times faster than the U.S. average. About 60,000 residents of Chattanooga (population 171,000) and 4,500 businesses subscribe to The Gig at prices that range from $29-$100 and up per month. The $29/month fee is targeted at low-income families with children, those who qualify for food stamps. This eliminates the homework divide we talked about earlier, enabling all children, regardless of means, to access the internet.

The Gig didn't just happen. It took a push from a forward-looking leader, Mayor Andy Berke, to make it happen plus federal funds. Today, The Gig is credited with an economic/entrepreneurial revival in Chattanooga with the following results:

→ **Three very large employers**[12] have moved in and set up shop: Alstom (which makes turbines for power plants), Amazon (with a distribution center bigger than 17 football fields), and VW (which employs 2,500 workers in an auto plant that cost $1B to build).

→ Chattanooga has gone from practically zero venture capital in 2009 to more than five organized funds with investable capital over $50M in 2014. (Source: **The Guardian, 2014**[13]).

→ A flourishing ecosystem exists to encourage entrepreneurial activity. (Source: **Ewing Marion Kauffman Foundation, 2016**)[14]

→ Retail and real estate are both taking off, experiencing double-digit growth (Source: **Wall Street Journal, 2012**[15])

→ Employment is up and unemployment is down. The population is growing.

The Ewing Marion Kauffman Foundation, the Obama Administration, and city government all credit The Gig with serving as a catalyst for economic development for Chattanooga; not too shabby for the city once called the dirtiest in America!

Still, municipal broadband is not without its challengers. Several states outlaw municipal broadband, reflecting the broad power of the telecom lobby. In response, the Obama administration authored a detailed report and set aside $7B for cities and metro-politan areas to make broadband available as a municipal service.

With broadband connectivity in place, citizens in the Maker city are able to advocate for specific projects that they know will make a difference in their city.

ENGAGING STUDENTS IN CIVIC PARTICIPATION

The key to civic innovation is engaging more people in the actual task of addressing city problems, and students are among its greatest resources. As we have seen in the education chapter of this Maker City book, when students are touched by the tools of Making they light up and can become deeply engaged in projects. The same is true when students are exposed to, asked to learn about, and solve real problems in the city. It's a twenty-first century approach to civic education where the city becomes a living lab. Contrast that with twentieth century civics (now seldom taught), which was a famously static affair: students were taught the branches of government or how a bill is passed, which is tough to directly relate to. In today's approach, civics can be a verb: it is something you do, and then often see results.

Citizen Schools is a national organization that engages mentors to work with at-risk urban middle school students in after school programs. The act of pairing students with professionals who are in technology, manufacturing, or finance is in itself enlightening and confidence building. One of the Citizen Schools' programs engages kids in the process of city planning, providing an apprenticeship in the architectural design process.

During the ten-week program students are introduced to the basics of urban planning and its connection to the social life of communities, taught to conduct ethnographic research, and shown how to use that research to formulate and design plans. Final presentations are delivered in front of urban planners and city officials. One can be excused for confusing this ten-week plan for middle schoolers with a graduate exercise in urban studies!

We expand a middle school's learning day

by connecting a team of adults

to provide relevant learning experiences

that give students the skills, access, and beliefs

they need to succeed in school, college, and careers.

Source: Citizen School

Apprenticeships have included designing a neighborhood, planning a community center, designing a playground, and reimagining a Boston MTA station. There is a full 100-page curriculum and how-to manual available from the **Citizen Schools website**[16].

Generation Citizen

Real world urban problems can be turned into student-led solutions. That's the mission of Generation Citizen, a national organization that takes the concept of civics and turns it into action projects where students understand, propose solutions, and act on their world.

Students find very real problems that affect their lives, problems they might otherwise feel no agency over, then research options, come up with solutions, and ultimately present their ideas to leaders to lobby for and supervise change.

For example, urban water quality has been in the news, with the crisis in Flint, Michigan illustrating the very real implications of water infrastructure decisions. Students at the Abraham Lincoln School Green Academy in San Francisco noticed a problem with water at their own school. Their campus recorded the second-highest water usage in the entire school district due in large part to an outdated infrastructure and a non-turf football field. However, students also weren't drinking from the school's dirty fountains, opting instead to buy sugary beverages or plastic water bottles.

Students decided to take multi-pronged action on water conservation at their school. They convinced the SF Public Utilities Commission to come to their campus to conduct water testing of toilets, sinks, and fountains. They were successful in getting new automatic sinks installed in all school restrooms from the school district, and testified at a SFUSD Board of Education meeting to demand more water filter stations and updated water infrastructure in schools district-wide. They wrote grants for water catch systems and improved appliances for their campus and met with Supervisor Katy Tang to request funding support at the district level.

In a twist to the traditional hackathon, Generation Citizen's Civic Tech Challenge pairs students with young adult software and hardware developers to address civic problems. The students specify the problems and act as "clients" with the hackers building out solutions and mentoring students. In Boston, in November 2015, teams addressed data visualization to reduce youth home-lessness, and developed several projects to take on the issue of gun violence through petitions, online social media, and a take action website. Using a classic web awareness/conversion/action model, students studied how they might have a real voice, then worked with hack teams to make sure legislators and the Mayor heard their stories and proposals. Because students cannot vote, often they do not get heard by city officials, teaching them at an early age that their default mode is to be disengaged from civic life. Not so with Makers!

These techniques can be practiced by any youth organization: schools, museums, boys and girls clubs, and more. What they have in common is bringing a Maker mentality to the civic realm.

Civics was once a static description of how government works. No longer.

Today civics can be about action, engagement, and creation: a set of active steps to research and deploy thoughtful prototype solu-tions. Students can apply a holistic approach to learning about a city's problems, engaging with mentors about solutions and formulating an approach using today's tools of Making and coding. This teaches civic responsibility and conveys a powerful sense of agency (I have control over my world!) that is essential for raising a generation of lifelong learners, Maker-capable citizens who can be productive in our coming economy and help reshape their world.

Source: Generation Citizen

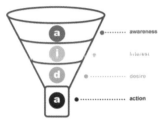

Classic model showing movement from awareness to action

IMPLICATIONS FOR CITIES

To open up a city to citizen engagement, Maker Cities are creating programs to:

 Open up their data. It turns out that open data is an invitation for citizen engagement in the city.

 Create opportunities for experimental programs on the streets of the city—what urban planning types call "tactical urbanism."

 Use technology to engage citizens who might otherwise be left behind in shaping the Maker City.

 Continually experiment through hackathons and festivals.

 Scale experiments through partnerships with academia, private companies, and public agencies. In other words, form cross-sector partnerships that insert Makers inside more established and establishment organizations.

 Start young! Engaged citizens are not born that way but need to be nurtured from a young age. Get Makers involved early, at the stage of a project where previously you might have involved a design-build architecture firm or star architect.

RESOURCE   QUICK WIN   IMPLEMENTATION ADVICE   POLICY   BIG IDEA

THE FUTURE OF
THE MAKER CITY

THOUGHTS ABOUT THE FUTURE OF OUR CITIES

The realization of Maker Cities portends the possibilities of a brighter future for our cities and their residents. In previous chapters, we looked at:

1. Characteristics of a Maker City

2. An introduction to the Maker mindset, Maker ethos, Maker tools

3. Ecosystems of Maker Cities

4. The Maker City as a learning community; what that means for K-12 education as well as lifelong learning

5. The challenges and promises of workforce and economic development in a Maker City

6. The renewal in urban manufacturing, especial advanced forms, inside the Maker City

7. Real Estate as a strategy to drive innovation in the Maker City

8. How a Maker City encourages Civic Engagement

There are implications for cities in each chapter that we hope are of use to individuals and organizations as they build out and strengthen the Maker City. The Maker City movement is still young, with much room for experimentation and growth.

What makes the Maker movement a movement is that much change is happening through the grassroot efforts of individuals and organizations that are committed to changing our cities for the better. This makes it more difficult (of course) to expose policy levers. You cannot legislate or zone a Maker City into existence. To move forward to strengthen the Maker City, city leaders and activists have the following building blocks they can work with:

→ **Cross-sector collaboration.** There is much evidence that when the public, private, and nonprofit sectors truly engage to solve a city's problems, the results can be transformative. Examples include: The Brooklyn Navy Yard, experiments in municipal broadband in Chattanooga, Tennessee; the collaboration of 250 organizations to remake learning throughout Pittsburgh, Pennsylvania.

→ **Co-creation, crowdsourcing, and rapid prototyping** are creating entirely new types of products that can be manufactured inside the Maker City. Examples include: consumer appliances (FirstBuild, Nomiku Wifi Sous Vide), fashion and textiles (Crye Precision, Manufacture NY), and precision components (energy production, aerospace).

→ **Engage major research universities to work with Makers.** Examples include: Carbon 3D, an advanced form of 3D printer that came out of Carnegie Mellon University and research into advanced material that—with funding from the Obama administration—is happening in collaboration with FIT and MIT.

→ **Engage with the Make Schools Alliance.** Seventy-eight colleges and universities in 32 states representing more than 1.1 million students have made commitments to support the Maker movement on their campuses and in their communities. These include the creation of Makerspaces open to students and the community, Maker training and certification, and mentorship to support local schools.

→ Embrace and support urban manufacturing in all shapes and forms but particularly advanced manufacturing as discussed in Chapter 6. The Urban Manufacturing Alliance is available to help you get experiments off the ground and achieve scale.

→ **Think small. Think local,** at least at first, to get experiments off the ground. Only later, when you know something works, does it makes sense to scale it up, by expanding your reach regionally and/or nationally. Examples include: Pittsburgh Children's Museum and Manufacture NY.

→ **Enable Makers to co-create public space.** Large civic art and renewal projects are giving way to smaller, more guerrilla efforts. Examples include urban prototyping and parklets in San Francisco and Makers working to rebuild Detroit in a phoenix-like fashion, taking back one blighted home at a time.

→ **Tap into young people as champions of change.** Examples include Ethan Toth of Wenatchee Washington who introduced his entire city to the power of Making and young people working with Fictiv and FirstBuild to manufacture precision parts. This isn't child labor; it's child capital, enabling young people to do productive work at an earlier point in their lives so as to build a lasting sense of agency and competency.

We know that the paths to Maker City status will vary, both in practice and timing, but given the speed at which these changes are taking place, we thought it prudent to use this chapter to look at the future of the Maker City.

The Institute for the Future (IFTF) is a think tank based in Palo Alto that has been researching and providing insight into the Maker movement and Maker Cities since 2008. We asked Rod Falcon, Research Director at Institute for the Future, to help imagine what the city of the future might look like thanks to the Maker City movement.

FROM MAKER CITIES TO OPEN CITIES

HOW THE MAKER MINDSET AND TECHNOLOGY ARE REINVENTING URBAN LIFE

BY ROD FALCON

Research Director, Institute for the Future

What is the future of the city? And how can you participate in making it? We can all start by cultivating a Maker mindset.

The Institute for the Future, a Palo Alto, California research nonprofit looked at the Maker mindset back in 2008. Back then, we forecasted how an emerging do-it-yourself culture of "Makers" would transform how goods, services, and experiences would be designed, manufactured, and distributed. Today, Makers represent a new kind of citizen, bringing creativity and inventiveness to urban life, opening up participation, resources, imagination, spaces, and economic opportunity. These Makers are quite literally fabricating a new kind of city: the Open City.

At the Institute for the Future, we explore the future of cities by looking at creators—the Makers and the communities that drive advances in technology, health, food, and civic engagement. We look at context, the environments that shape and are shaped by technology and that amplify or disrupt people's lives. We look at home, work, and the streets where life unfolds. We also look at technologies, the tools that drive new ways of interacting with things and with each other.

By exploring these we see early signals of how the city is being remade. Cities are becoming places where people and new tools are coming together in transformative ways. Cities are our laboratories for the future. They're where we face the future first. They're where people and ideas from around the world come together in a massive, unpredictable, crucible of innovation.

Today, the world's cities remain places of creativity and experimentation. More people worldwide live in cities than those who do not. But they're also places that face challenges.

We see that our transportation, food, health, education, and governance systems are unprepared for the scale of change that's going to take place over the next decades. This means that the ways we've built infrastructure, provisioned services, and coordinated whole economies and markets need reinvention. So, how will we meet these challenges?

Science-fiction author William Gibson has some answers for us.

"The future is here, it's just not evenly distributed."

This quote of his is really fundamental in that it teaches us to look for the future in the present.

But there is another William Gibson quote that gives us a hint of where in the present we should look, if we want to find the future.

"The street finds its own uses for things."

His insight is that whatever a technology is designed for, people will find their own ways to use it. That's what the Maker mindset is about. And we've found that people all over the world have adopted this mindset that questions how we're intended to use technologies. And in doing so, they generate new innovations, new uses, and sometimes complete transformations, true break-throughs from the way we used to do things.

If we look at the Maker movement and the number of Maker Faires happening across the globe, you can see how big and global this is. But this is just the most visible example of a mindset that is more widespread. Even before anyone called it that, the Maker mindset thrived in the streets. For example, take the turntable. The turntable was designed as a technology to play music. But not everyone on the street saw it that way. In the Bronx in the 1970s, young people generally didn't have bands at dance parties. Bands were expensive. But what people did have was the turntable.

And just the way a band feels out the room and adjusts how they play, so does a DJ. They read the room and choose what record to play next. But one DJ, a Jamaican immigrant who went by the name DJ Kool Herc, took it one step further.

He listened to the music carefully and watched the crowd. And he noticed that there was one part of every song that drove people on the dance floor crazy. And he had this idea, what if you could use the turntable to isolate the breaks and put them all together:

"Hmm... They're waiting for this one particular break and I have a couple more records that have the same break up in it. I wonder how would it be if I put them all together. I started out with James Brown, 'clap ya' hands, stomp ya' feet.' And that part right there, the break, I'd come in with Apache, 'Bongo Rock.'"

DJ Kool Herc in NYC May 1999 By mika-photography.com via Wikimedia Commons

And that was the birth of hip-hop. His story is about the Maker mindset in action. Herc took something that was designed to be a technology of consumption and turned it into a technology for creation. And he helped create the future of popular music worldwide.

Today we're beginning to see that the Maker mindset is not just about making or repurposing stuff but remaking systems, those urban systems that need reinvention. The city itself is the new turntable.

In Shenzhen, China we met Makers who are resetting technology manufacturing to super high speeds in Hua Qiang Bei. And we experienced a Maker City that is getting reinvented in the Chinese imagination as a frontier for experimentation and innovation.

In Detroit, we saw citizen resourcefulness in action as communities build their own services despite the collapse of institutions around them.

We immersed ourselves in Mexico City, where the Maker movement is just emerging in this megacity, but we saw first-hand how it's being embraced rapidly as a way to get things done without a lot of resources.

We learned that Maker Cities are becoming Open Cities. Open Cities reflect the value of accessibility through the strategies of participation, sharing, imagination, adaptability, and equity. And in city after city, Makers shared similar values, embracing openness and experimentation, hacking not only technology and innovation challenges but urgent urban challenges as well. Out of their experiments will emerge a core set of Open City strategies that will define the urban experience for decades to come.

Think of these strategies as a manifesto for open cities. They expand what is possible by provoking us to ask questions such as:

Early iPhone Application: "Five O" or "Yelp for Police"

1. What if cities could expand participation?

The idea behind this strategy is how to let more people participate in commercial and civic life. Think about how Yelp works. Yelp depends on participation. It needs users to provide feedback, and by harnessing participation it created one of the most popular sources of reviews for restaurants and other businesses. This is an example where participation creates new commercial value, but we can also see citizens start to use the same strategies—the same kind of platforms—to create new forms of civic participation. Take for instance teenagers in Georgia who created their own Yelp for police. They call it "Five O". Frustrated by a lack of accountability, these kids invented an app that allows users to document interactions with police officers and rate their behavior. Here we can see the Maker mindset is not just about repurposing devices, but entire platforms, businesses, and engagement models for community needs.

2. What if cities were designed for sharing?

This Open City strategy looks for underutilized assets and unlocks capacity in people, places, and things. Cities have enormous latent capacity enabling citizens to share tangible goods, raw data, expertise, time, or assistance. Coordination tools for sharing excess capacity allow us to extract more value from people, places, and things and create new kinds of commerce as well as charitable giving. At the same time, apps like Waze, which aggregate user data to create traffic maps, point to new ways that sharing citizen-generated data will create entirely new experiences in the urban landscape by generating new value and illuminating how the city can work better for everyone.

3. What if cities inspired and harnessed imagination?

Back in 2013 San Francisco and the Make-A-Wish Foundation showed us what happens when we empower people by inspiring their imaginations. When the foundation decided to grant a boy his wish to be "Batkid" for a day, they anticipated a few hundred people would pitch in to make it happen. But the whole city got involved. Businesses, politicians, and more than 10,000 citizens all coordinated to transform San Francisco into Gotham City. The local newspaper even put out a Gotham City Examiner. All this wonderful energy and imagination came together for one child for one day. But what if we could harness the same energy and imagination everyday to remaking the city? The next decade's virtual reality and augmented reality tools will enable artists and citizens to reimagine their communities and persuade others through immersion in "what-could-be" scenarios. As lightweight tools for simulation and prototyping emerge, anyone will be able to create and share visions for the future of the city.

4. What if cities made public spaces adaptable?

Today, food truck courts and parklets—those curbside parking spaces converted to public benches and walkways—are among the most mainstream examples of a broader trend toward recolonizing urban spaces to make them more open, public, and social. The shifting demographics of cities will transform needs, habits, and even values, which in turn will change the demand and priorities for the use of public spaces. At the same time, the advent of lightweight manufacturing technologies and crowd-sourced "recipes" for doing almost anything will accelerate the ability to place-make. With more adaptable spaces, cities will become more open to the needs and imaginations of populations.

5. What if cities created more equity across the population?

A city can use all of the other Open City strategies but do little to address equity. At their core, Open Cities are about creating a new culture of innovation that pursues equity by making spaces, services, and economic opportunities open and accessible to all citizens regardless of age, ability, gender, and socioeconomic status—that is, they're about creating new standards of equity. As they leverage open data, shareable resources, and adaptable spaces to meet emerging needs across diverse populations, Open Cities create alternative pathways to attain equity.

MAKE THE FUTURE TODAY

Each urban challenge we face can't be solved by any one stake-holder. But the promise of Open Cities is that we can enlist each other. Together we can untangle the knotty problems that ensnare communities and cities. The future is within reach. You can start making the future by cultivating your Maker mindset and partici-pating in the reinvention of your city as an Open City. IFTF's, Open Cities: How Technology and the Maker Mindset is Reinventing Urban Life, is a Maker kit for doing just that. The map contains tools and a process to help you choose your Open City strategies. It highlights the technology catalysts Makers are using to accel-erate their strategies. It shares how Makers are creating new tools and platforms to meet existing and emerging needs and in the process remaking how we all live, work, and play.

Open Cities is a big idea. It inspires and overwhelms at the same time. Our hope is that we can help you reimagine your city or any city as an Open City and think about ways to bring the future forward.

After all, we're all invested in the future of our cities. We invite you to explore the future, to imagine the possibilities, and to start making the future today.

IMPLICATIONS FOR CITIES

Download the IFTF research map and toolkit on **Open Cities** to gain additional insight on how to better anticipate future trends that will affect the evolution of your city, from Maker City to Open City and beyond.

IFTF Open Cities Map
mcbook.me/2cykb50

Anticipate the future by making the five strategies listed above your own:

1. What if my city could expand participation?
2. What if my city was designed for sharing?
3. What if my city inspired and harnessed imagination?
4. What if my city made public spaces adaptable?
5. What if my city created more equity across the population?

Follow all or some of the people and organizations from our twitter **lists** to learn more about Maker Cities.

TWITTER LIST
mcbook.me/twitterlist200

Contribute your own story to the Maker City project, by authoring a case study on Medium.com and using the hashtag #makercity when you post.

Work with a local college or university to field an economic impact study and share the results on Medium using that same hashtag.

If there are policy hacks, funding sources, or recent legislation that Maker Cities could benefit from, let us know and we'll get the word out.

Sponsor a meetup, Maker Faire, and/or Makerspace (hint: crowdsource to get your initial financing) if one does not already exist in your city.

Build the ecosystem you need, one that ideally cuts across sectors, including the learning network for K-12 education that already exists in your city, colleges and research universities, as well as the startup/ entrepreneurial community and innovation centers within major corporations.

 Invest in urban manufacturing by partnering the Urban Manufacturing Alliance, an organization uniquely qualified to help you put in place programs to encourage urban manufacturing in your city.

 Support young people who want to go into Making. After 40 years of vilifying people who work with their hands, it's important that we tell young people with a background in STEM that they can support themselves and their families by working with their hands. The next Elon Musk or Steve Jobs is almost certainly working away today as a master carpenter, a robotics maker, a drone designer, and/or fabricator of prosthetics or another kind of custom medical device.

 Support older workers as they retire and guide them into roles where they can transfer the knowledge they have as well as their mastery of specific Maker skills to others.

 Identify and nurture new forms of business. Today, it is possible to put the means of production into more people's hands at a lower cost than at any other time in history.

 Remember that career paths inside corporation are— for all intents and purposes—gone. Encourage young people to build a portfolio of relevant projects they have completed and also to get training in the "Understandings" of starting their own businesses based on their Maker skills.

RESOURCE   QUICK WIN   IMPLEMENTATION ADVICE   POLICY   BIG IDEA

CHAPTER 1

1. White House Initiative (2014) - mcbook.me/2cqDjV7

2. Open Innovation site - openinnovation.net

3. Fast Company Design (2015) - mcbook.me/2cMgg6M

4. Boston Globe (2015) - mcbook.me/2c6Zalh

5. Economic Progress RI (2014) - mcbook.me/2c4MVmh

6. NAHPY Awards (2012) - mcbook.me/2cMhX3X

7. Cities Alliance frameworks - mcbook.me/2ci1gJs

8. Kauffman Foundation Blog (2015) - mcbook.me/2cAYdNh

9. Kauffman Index of Entrepreneurship Series - mcbook.me/2bYODTa

CHAPTER 2

1. Maker Faire website - mcbook.me/2cdRzvP

CHAPTER 3

1. Deloitte University Press (2014) - mcbook.me/2cB26SB

2. American Library Association site - mcbook.me/2c4Qid1

3. FabLab Foundation site - mcbook.me/2ccVXg1

4. Artisan's Asylum (2014) - mcbook.me/2cglPIO

5. Seeed Studio (2013) - mcbook.me/2c4QetD

6. Portland Made Collective survey (2014) - mcbook.me/2ccXyCk

7. Portland Made Collective survey (2015) - mcbook.me/2cMpZep

8. Sample Maker survey available on last 2 pages of this report - mcbook.me/2ccXyCk

CHAPTER 4

1. Education 101 class on Dewey - mcbook.me/2ckBEig

2. Simply Psychology on Piaget - mcbook.me/2cMqVzD

3. Stanford University on Papert - mcbook.me/2cdUG79

4. New York Times interactive feature (2012) - mcbook.me/2bYSxeB

5. Edutopia.com on Project-Based Learning - mcbook.me/2cdUPY5

6. MIT admissions site on how to submit your Maker portfolio - mcbook.me/2cMqAwx

7. Mayor's resolution on digital badges (2016) - mcbook.me/2cMqjtr

8. Allen Touch academic paper on informal adult learning (2002) - mcbook.me/2cgomCM

9. White House Mayor's Maker Challenge (2014) - mcbook.me/2bYU0BE

10. LRNG handbook for partners (2016) - mcbook.me/2bYT5Rh

CHAPTER 5

1. Daily Beast on revival in the rustbelt (2014) - mcbook.me/2cBvc4i

2. Governing on attracting a millennial workforce (2015) - mcbook.me/2cehnYP

3. Freelance Union Elance/Odesk Study (2014) - mcbook.me/2cMTD1K

4. Etsy Study (2015) - mcbook.me/2cgKI6V

5. Etsy blog post (2013) - mcbook.me/2c7C65Y

6. CNC West (2016) - mcbook.me/2cfIGU0

7. Georgetown Center on Education and the Workforce report (2014) - mcbook.me/2c5ekVn

8. California Council on Science and Technology (2016) - mcbook.me/2c7CNMj

9. Urban Institute (2007) - mcbook.me/2cMM3po

10. Hozler & Lerman, Brookings on middle skill jobs (2009) - mcbook.me/2cgKCwf

11. GOA report on Workforce Funding - mcbook.me/2cehIdY

12. Harvard Business Review on the skill gap (2014) - mcbook.me/2c5fNe0

CHAPTER 6

1. Brookings Institute on Rustbelts Global Innovation (2016) - mcbook.me/2cl0MFE

2. Deloitte Center for the Edge on Future of Urban Manufacturing (2015) - mcbook.me/2csPMq6

3. Harvard Business Review on Clusters and New Economics of Competition (1998) - mcbook.me/2ciJrda

4. Deloitte Advanced Manufacturing Technologies report - mcbook.me/2cMYP5O

5. Route Fifty on Fremont BART extension (2016) - mcbook.me/2crbQCE

6. Mercury News on Kraft Heinz job cutbacks (2015) - mcbook.me/2cy85su

7. Inside EV on Tesla Model 3 (2016) - mcbook.me/2ceiiIF

8. Small Business Administration Growth Accelerator Fund - mcbook.me/2bZiv1o

9. Pratt Center Nonprofit Real Estate toolkit - mcbook.me/2cKyLW4

10. McKinsey on Future of Manufacturing (2012) - mcbook.me/2ceiNCx

11. Brookings on Urban Manufacturing (2011) - mcbook.me/2cdIEfe

CHAPTER 7

1. Pratt Center Brooklyn Navy Yard Report (2013) - mcbook.me/2cKzOoY

CHAPTER 8

1. Insititute for the Future on Open Cities (2014) - mcbook.me/2cMNKTH

2. Christian Science Monitor on tiny homes (2016) - mcbook.me/2c0Rv6I

3. Curbed on mapping San Francisco's Parklets (2014) - mcbook.me/2cN6voS

4. PBS on Robert Moses (2013) - mcbook.me/2bZlffl

5. Gizmodo Interview with Blaine Merker Gehl Studio (2013) - mcbook.me/2crhyEC

6. Gehl Architects report on Market Street Prototyping Festival (2015) - mcbook.me/2cBBUqQ

7. Instructables Market Street Prototyping - mcbook.me/2ciN5Un

8. Socrata on running a hackathon - mcbook.me/2csTpwa

9. Data Canvas on do-it-yourself sensors - mcbook.me/2crjGwq

10. New York Times on effects of digital divide on school children (2016) - mcbook.me/2c7JPAN

11. City Lab on New York Broadband for All (2015) - mcbook.me/2c0TpUN

12. Wall Street Journal on Chattanooga, Tenn. (2012) - mcbook.me/2cgOWLX

13. The Guardian Chattonaooga Gig internet tech boom (2014) - mcbook.me/2c0SKmA

14. The Ewing Marion Kauffman Foundation Case Study of Chattanooga (2016) - mcbook.me/2cNaOk6

15. Wall Street Journal on Chattanooga, Tenn. (2012) - mcbook.me/2cgOWLX

16. Citizen Schools how-to manual on apprenticeships - mcbook.me/2c5iUmy

CHAPTER 9

1. Institute for the Future Open Cities Map - mcbook.me/2cykb50

The Maker City book is designed to help cities understand the Maker movement and the impact it is having on economic opportunity: ecosystem development, education, advanced manufacturing, workforce development, and real estate.

The book is based on interviews completed with 50 thought leaders and practitioners who work at the forefront of the Maker movement in around 20+ cities including:

Brooklyn, New York

Chattanooga, Tennessee

Chicago, Illinois

Cleveland, Ohio

Detroit, Michigan

Elyria, Ohio

Fremont, California

Ithaca, New York

Kansas City, Missouri

Lewisburg, Pennsylvania

Los Angeles, California

Louisville, Kentucky

Macon, Georgia

New York, New York

Pittsburgh, Pennsylvania

Portland, Oregon

Providence, Rhode Island

Raleigh, North Carolina

San Diego, California

San Francisco, California

Wenatchee, Washington

To learn more about Maker Cities, please visit makercity.com.

This book would not have been possible without the generous financial support of the Ewing Marion Kauffman Foundation. Additional support provided by:

CITY INNOVATE
FOUNDATION

GRAY AREA

INSTITUTE FOR THE FUTURE

Ewing Marion
KAUFFMAN
Foundation

MAKER MEDIA

Urban
Manufacturing
Alliance

EXECUTIVE OFFICE OF THE PRESIDENT OF THE UNITED STATES

Interested in adding your own story to the mix? We invite you to do so by blogging on Medium and Twitter using the hash tag #makercity.

The Maker City book was made possible by the generous financial support of the Ewing Marion Kauffman Foundation (kauffman.org) with fiscal sponsorship provided through the Gray Area Foundation for the Arts (grayarea.org).

Stephanie Santoso, formerly Senior Advisor for Making to the White House Office of Science and Technology Policy (OSTP), Executive Office of the President, was invaluable in providing introductions to people who make up the Maker City ecosystem. Stephanie has moved on to a new position in California and recently got married; in the face of the significant changes in her life, we cannot thank her enough for remaining with us to the very end of this project.

An online version of these acknowledgement provides links to the organizations and individuals listed here and can be found at makercity.com/acknowledgments.

The inspiration for the Maker City book came very much from the Remake Learning Playbook produced by the Sprout Fund under the leadership of Cathy Lewis Long, President.

Interviews

In all we did about 50 interviews with thought leaders, practitioners, movers and shakers across ~20 cities. These people donated their time, thoughts, and energies to helping us understand what made their Maker City tick. We cannot thank them enough.

Baltimore, Maryland
Andrew Coy, Executive Director, Digital Harbor Foundation

Brooklyn, New York
Bob Bland, CEO & Founder, Manufacture New York

Matthew Burnett, CEO & Co-Founder, Makers Row

David Ehrenburg, CEO & Executive Director, Brooklyn Navy Yard Development Corp.

Adam Friedman, Executive Director, Pratt Center for Community Development

Chattanooga, Tennessee
Mike Bradshaw, Founder & CEO, Company Lab (Co.Lab)

Cleveland, Ohio
Lisa Camp, Dean of Strategic Initiatives, Case Western University

CJ Lynce, TechCentral Manager, Cleveland Public Library

Sonya Pryor-Jones, Chief Implementation Officer, Fab Foundation

Felton Thomas, Director, Cleveland Public Library

Detroit, Michigan
Carol Colletta, Senior Fellow American City's Practice, The Kresge Foundation.

When we interviewed Ms. Colletta she held a similar position at The Knight Foundation

Bill Coughlin, President & CEO, Ford Global Technologies LLC

Jeff Sturges, Founder, Mt. Elliott Makerspace

Jen Guarino, VP of Leather Goods, Shinola

Elyria, Ohio
Kelly Zelesnik, Academic Dean, Engineering , Business & Information Technologies and Nord Advanced Technologies Center, Lorain County Community College

Fremont, California
Kelly Kline, Economic Development Director/Chief Innovation Officer, City of Fremont

Ithaca, New York
David Schneider, Senior Lecturer, Systems Engineering, Cornell University

Kansas City, Missouri
Aaron Deacon, Managing Director, KC Digital Drive

Lewisburg, Pennsylvania
Margot Vigeant, Associate Dean, College of Engineering, Bucknell University

Los Angeles, California
Krisztina 'Z' Holly, Founder, Make it in LA, an initiative of City of Los Angeles Mayor's Office

Louisville, Kentucky
Danielle Blank, President, LVL1 Hackerspace

Ted Smith, Chief of Civic Innovation, City of Louisville

Natarajan "Venkat" Venkatakrishnan, Director of Research and Development for GE Appliances, FirstBuild

Macon, Georgia
Nadia Osman, Director, SparkMacon

New York, NY
Anthony Townsend, Sr. Research Scientist, NYU Rudin Center for Transportation Policy and Management

Luke Dubois, Professor of Integrated Digital Media, NYU

Pittsburgh, Pennsylvania
Gregg Behr, Executive Director, Grable Foundation

Daragh Byrne, Co-Lead, Make Schools Alliance/Carnegie Mellon

Sunanna Chaud, Learning Innovation Strategist, Remake Learning

Bernie Lynch, Project Manager, The New App for Making It In America, a project funded by the Institute for Work & the Economy (Chicago, Illinois)

Timothy McNulty, Associate Vice President for Government Relations, Carnegie Mellon University

Anne Sekula, Director, Remake Learning Council

Jane Werner, Director, Children's Museum of Pittsburgh

Providence, Rhode Island
Bert Crenca, Founder/Plenipotentiary, AS220 Maker Space

Raleigh, North Carolina
Ellen Chilton, Merchant Promotions Manager, Downtown Raleigh Alliance

Aly Khalifa, Founder, LYF Shoes & Designbox

Sean Maroni, Former Fellow, Betabox

Riverside, California
Gene Sherman, CEO & Founder, Vocademy Makerspace

San Diego, California
Susie Armstrong Senior Vice President, Engineering, Qualcomm

Erin Gavin, Senior Manager, Government Affairs, Qualcomm

Travis Good, Organizer, San Diego Maker Faire

Misty Jones, Director, San Diego Central Library

San Francisco, California
Deborah Cullinan, CEO, Yerba Buena Center for the Arts (YBCA)

Dave Evans, Founder & CEO, Fictiv

Mark Hatch, Former CEO & Co-Founder, TechShop

Steven Heintz, VP of Engineering and General Manager, Flex Innovation Labs

Neil Hrushowy, Manager, City Design Group at San Francisco Planning Department

Madelynn Martiniere, Director of Community, Fictiv

Patrick Dunne, Director of Industrial Applications, 3D Systems

Claire Micheals, Manufacturing Workforce and Hiring Manager, SFMade

Nick Pinkston, CEO & Founder, Plethora

Kate Sofis, Executive Director, SFMade

Wenatchee, Washington
Steve King, Director, Community & Economic Development Department, City of Wenatchee

Karen Rutherford, Council Member, Wenatchee City Council

Ethan Toth, Student/Coordinator, Maker Faire Wenatchee

Allison Williams, Executive Services Director, City of Wenatchee

• • •

Contributors

Kim-Mai Cutler, Reporter, TechCrunch

Rod Falcon, Program Director, Technology Horizons, Institute for the Future

Adam Friedman, Executive Director, Pratt Center for Community Development

Thomas Kalil, Deputy Director for Policy for the White House, Office of Science and Technology Policy, Executive Office of the President; Senior Advisor for Science, Technology and Innovation for the National Economic Council

Heather King, Consultant and former Managing Director, Eight Inc.

Josette Melchor, Executive Director and Founder, Gray Area Foundation for the Arts

Mark Muro, Senior Fellow and Policy Director, Metropolitan Policy Program Brookings Institute

Jenn Sander, Global Initiatives Advisor, Burning Man Project

Stephanie Santoso, Senior Advisor for Making at White House, Office of Science and Technology Executive Office of the President

Lauren Sinreich, Lead Researcher and Contributor

* * *

Reviewers

Andrew Coy, Senior Advisor at White House, Office of Science and Technology Policy, Executive Office of the President

Penelope Alice Douglas, Innovation Consultant

Mark Hatch, Former CEO & Founder, TechShop

Mickey McManus, Chairman and Principal, Maya Design

Jason Wiens, Policy Director, Ewing Marion Kauffman Foundation

* * *

Production

Mary F. Doan, Executive Producer

Omeed Manocheri, Online Producer

* * *

Graphic Design

Erik van der Molen, Principal, Office vs. Office

Peter Hirshberg

Peter Hirshberg serves as an innovation advisor to cities and companies. He has created two centers of urban innovation from scratch: City Innovate Foundation, an organization formed with The San Francisco Mayor's office, UC Berkeley, and the MIT Media Lab; and Gray Area Foundation for the Arts, one of San Francisco's most vibrant and civic-minded arts organizations, a center for digital media education, incubation, performance, and exhibition.

He is CEO and founder of the Re:Imagine Group helping brands with strategy and marketing in a world of empowered connected audiences. He's worked with executive teams at Best Buy, Sony, IBM, Verizon, Time Warner, Unilever, GE, Estee Lauder, Telefonica, and many others on their innovation and digital growth strategies.

Previously, Hirshberg was Chairman of Technorati, the pioneering social media search engine and advertising network with over a billion monthly page views. During his nine-year tenure at Apple Computer, Hirshberg headed Enterprise Marketing, where he grew Apple's large business and government revenue to $1 billion annually.

Peter is a sought after technology and innovation speaker, having presented at TED, the World Economic Forum, DLD, The Aspen Ideas Festival, E.G., Techonomy, CeBIT, WEB 2.0 Summit, and many other events.

His board and advisor positions have included Active Video Networks, Technorati, Build Public, The Computer History Museum, and Gray Area Foundation for the Arts. He is a Senior Fellow at the USC Annenberg Center on Communication Leadership and Policy and a Henry Crown fellow of the Aspen Institute.

You can reach peter@makercitybook.com or on Linked In at linkedin.com/in/hirshberg

Dale Dougherty

Dale Dougherty is the CEO and Founder of Make Media, publisher of Make Magazine. He has been called the "godfather of the Maker movement," having founded both the magazine and the first Maker Faire in 2006. Dale was instrumental in launching the Maker movement which is reshaping science and technology education and creating unprecedented economic opportunity for young people and a renaissance in urban manufacturing.

Dale started down this path in 1984, when he worked with Tim O'Reilly to create O'Reilly Media, publisher of "Nutshell" handbooks. At O'Reilly Media Dale developed the Hacks series of books, coined the term Web 2.0, and then creating the Web 2.0 Conference, a joint venture between O'Reilly and CMP.

The idea for Make Magazine came from his experiences with the Hacks books, as Dale recognized that hackers were playing with hardware and more broadly, were looking at how to hack the world, not just computers. Make Magazine is a re-invention of Popular Mechanics for the 21st Century, and it launched in 2005. A year later, Dale launched the first Maker Faire in San Mateo to bring together all the Makers to share their work with each other and with the public. In 2016, there will be 150+ Maker Faires worldwide.

In addition to the Maker City book, Dale is the author of the upcoming book Free to Make (Fall 2016, North Atlantic Books) and sits on the boards of Make Media, Maker Education Initiative, and the nonprofit Big Picture Learning.

You can reach Dale at dale@makercitybook.com or on Linked In: linkedin.com/in/daledougherty

Marcia Kadanoff

Marcia Kadanoff is a civic innovator and innovation advisor to nonprofit and for-profit companies. She serves as Chief Strategy Officer at City Innovate Foundation and as a member of their board of directors. Prior to that she served as a business and marketing strategist, providing guidance to the C-Suite at 3Com, AOL, Content Rules, Levi Strauss, Microsoft, Oracle, Presence Learning, Charles Schwab, and Wells Fargo.

Previously, Marcia worked as a marketing executive at Apple Computer, as a VP of Marketing at Sun Microsystems, and co-founded the largest independent direct marketing agency on the West Coast.

She is a previous member of the Committee of 200 for high-echelon women leaders; a current member of Watermark which seeks to increase the representation of women on boards and as entrepreneurs and executives; and serves on the board of SF Camerawork. Marcia has a strong interest in equitable development inside our cities, having served on the board of Rebuilding Together SF and having worked as a communications advisor to the Tenderloin Neighborhood Development Corporation.

You can reach Marcia at marcia@makercitybook.com or on Linked In at linkedin.com/in/marciak

259

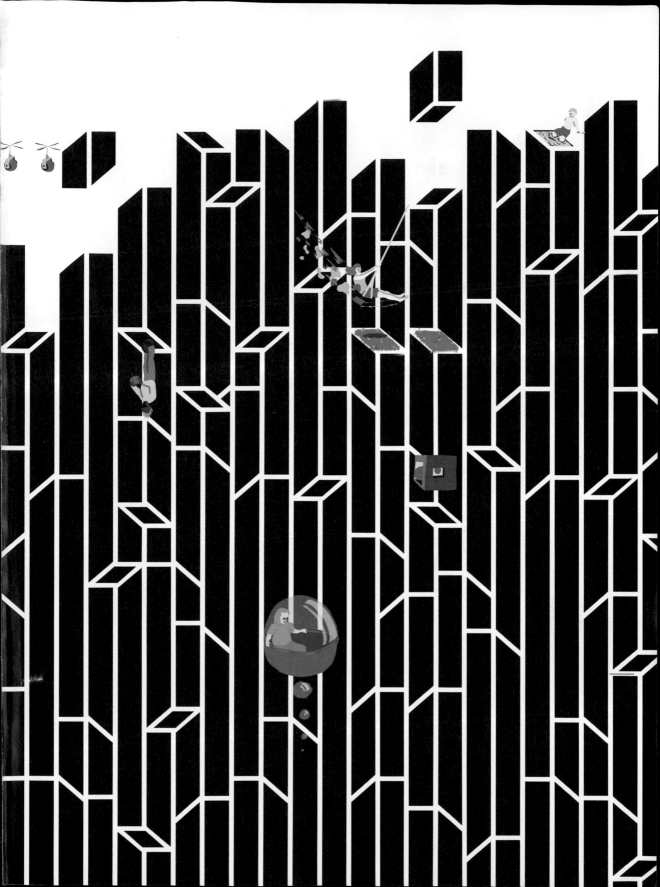